Collins

Edexcel Level 2 Extended Maths Certificate

Study and Practice

Anne Stothers
Andrew Milne
Trevor Senior

Introduction

Welcome to Collins Edexcel Level 2 Extended Maths Certificate Study and Practice Book. This book will help you to progress with ease through more challenging maths work, with detailed examples and plenty of practice in the key areas needed to succeed in maths at a higher level.

This book is suitable to support the Edexcel Level 2 Extended Mathematics Certificate. It covers the full specification content for this qualification, and can be used as a class text or self-study resource.

KEY POINTS AND WORKED EXAMPLES

Remember the important points of each topic with key words and succinct topic explanations. Gain a greater understanding of each area by studying the detailed worked examples before each exercise.

GRADED EXERCISES

Consolidate your knowledge of topics and know exactly what level you are working at, with hundreds of graded practice questions. Stretch and challenge yourself with questions targetting the four levels of grades, from Pass to Distinction*. Look for the non-calculator symbol for opportunities to practise your skills for the non-calculator assessment.

HINTS AND TIPS

Find valuable hints and tips boxes throughout the book that give you handy methods to try when answering questions.

EXAM PREPARATION

Ensure you are ready for your exams by tackling the exam-style questions at the end of each chapter.

ANSWERS

All answers to the practice exercises and exam-style questions in this book can be found at the back of the book.

Contents

1 Number 4
  1.1 Indices 4
  1.2 Surds 9
  1.3 Rationalising denominators 11

2 Algebraic manipulation 15
  2.1 Algebraic indices 15
  2.2 Expanding brackets 18
  2.3 Factorising 22
  2.4 Completing the square 30
  2.5 Algebraic fractions 32

3 Graphs 36
  3.1 Linear graphs 36
  3.2 Quadratic graphs 43
  3.3 Cubic and quartic graphs 50
  3.4 Trigonometric graphs 58

4 More graphs 64
  4.1 Translating and reflecting graphs 64
  4.2 Stretching graphs 70
  4.3 Circles 73
  4.4 Exponential and reciprocal graphs 77
  4.5 Non-linear graphs 84

5 Functions 98
  5.1 Functions 98
  5.2 Composite functions 104
  5.3 Inverse functions 107
  5.4 Transforming functions 110

6 Equations and inequalities 115
  6.1 Solve equations 115
  6.2 Solve quadratic equations 118
  6.3 Solve simultaneous equations 126
  6.4 Solving inequalities 132

7 Pythagoras' theorem and trigonometry 142
  7.1 Pythagoras' theorem in 2D and 3D 142
  7.2 Trigonometry in 2D and 3D 145
  7.3 Sine rule, cosine rule and area of a triangle in 2D and 3D 149

8 Probability 161
  8.1 The language of probability 161
  8.2 Conditional probability 166

9 Proof 173
  9.1 Proof by deduction 173
  9.2 Proof by exhaustion and disproof by counter example 177
  9.3 Geometric proofs 180

10 Vectors 186
  10.1 Position vectors 186
  10.2 Solving geometric problems 192

Answers 197

1 Number

1.1 Indices

THIS SECTION WILL SHOW YOU HOW TO ...
✓ multiply and divide expressions that contain indices
✓ simplify expressions involving negative indices, which may be written in a variety of forms
✓ simplify expressions involving fractional and negative indices, which may be written in a variety of forms

KEY WORDS
✓ index   ✓ indices   ✓ power
✓ negative index ✓ reciprocal

Rules of indices

To *multiply* **powers** of the same base number or variable, *add* the **indices**.

$3^4 \times 3^5 = 3^{(4+5)} = 3^9 \qquad 4^5 \times 4^{-2} = 4^{(5+(-2))} = 4^3 \qquad 2^3 \times 2^4 \times 2^5 = 2^{12}$

To *divide* powers of the same base number or variable, *subtract* the indices.

$3^6 \div 3^4 = 3^{(6-4)} = 3^2 \qquad 4^5 \div 4^{-2} = 4^{(5-(-2))} = 4^7 \qquad \dfrac{2^7}{2^3} = 2^4$

When you *raise* a power to a further power, you *multiply* the indices.

$(4^2)^3 = 4^{2 \times 3} = 4^6 \qquad (5^3)^4 = 5^{12} \qquad (2^{(-3)})^{(-4)} = 2^{12}$

EXERCISE 1A

TARGETING PASS

1. Write these as single powers.
   a) $5^2 \times 5^2$ b) 5×5^2 c) $5^2 \times 5^4$ d) $5^6 \times 5^{-3}$
   e) $5^2 \times 5^3$ f) $6^5 \div 6^2$ g) $6^4 \div 6$ h) $6^4 \div 6^2$
   i) $6^5 \div 6^{-2}$ j) $6^5 \div 6^4$

2. Write these as single powers of 4.
   a) $(4^2)^3$ b) $(4^3)^5$ c) $(4^1)^6$
   d) $(4^3)^2$ e) $(4^4)^2$ f) $(4^7)^0$

HINTS AND TIPS

Remember that the base numbers must be the same for index rules to apply.

Negative indices

A **negative index** is a convenient way of writing the **reciprocal** of a number or term. Here are some examples.

$3^{-1} = \dfrac{1}{3} \qquad 5^{-2} = \dfrac{1}{5^2} = \dfrac{1}{25} \qquad 2^{-3} = \dfrac{1}{2^3} = \dfrac{1}{8} \qquad \left(\dfrac{2}{3}\right)^{-1} = \dfrac{3}{2}$

If the base number you are evaluating is fractional, you can apply the power to both the numerator and the denominator.

$$\left(\frac{2}{3}\right)^{-2} = \left(\frac{3}{2}\right)^{2} = \frac{3^2}{2^2} = \frac{9}{4} \qquad \left(1\frac{1}{4}\right)^{-3} = \left(\frac{5}{4}\right)^{-3} = \left(\frac{4}{5}\right)^{3} = \frac{4^3}{5^3} = \frac{64}{125}$$

EXAMPLE 1

Rewrite the following in the form 2^n.

a) 8 b) $\frac{1}{4}$ c) -32 d) $-\frac{1}{64}$

a) $8 = 2 \times 2 \times 2 = 2^3$ b) $\frac{1}{4} = \frac{1}{2^2} = 2^{-2}$

c) $-32 = -2^5$ d) $-\frac{1}{64} = -\frac{1}{2^6} = -2^{-6}$

EXERCISE 1B

TARGETING PASS

1. Write each of these in fraction form, using indices.
   a) 5^{-3} b) 6^{-1} c) 10^{-5} d) 3^{-2} e) 8^{-2} f) 9^{-1}

2. Write each of these in negative index form.
   a) $\frac{1}{3^2}$ b) $\frac{1}{6}$ c) $\frac{1}{10^3}$

3. Change each expression into an index form of the type shown.

   a) All of the form 2^n
   i) 16 ii) $\frac{1}{2}$ iii) $\frac{1}{16}$ iv) -8

   b) All of the form 10^n
   i) 1000 ii) $\frac{1}{10}$ iii) $\frac{1}{100}$ iv) 1 million

   c) All of the form 5^n
   i) 125 ii) $\frac{1}{5}$ iii) $\frac{1}{25}$ iv) 1

   d) All of the form 3^n
   i) 9 ii) $\frac{1}{27}$ iii) 1 iv) -243

4. Evaluate the following, giving your answer as a fraction or mixed number.
   a) $\left(\frac{3}{7}\right)^{-2}$ b) $\left(\frac{2}{5}\right)^{-3}$ c) $\left(\frac{1}{4}\right)^{-1}$

Fractional indices (numerator = 1)

Consider the equation $7^x \times 7^x = 7$.

This can be written as $7^{(x + x)} = 7$
$$7^{2x} = 7^1$$
$$2x = 1$$
$$x = \tfrac{1}{2}$$

If you now substitute $x = \tfrac{1}{2}$ back into the original equation, you see that $7^{\tfrac{1}{2}} \times 7^{\tfrac{1}{2}} = 7$

So $7^{\tfrac{1}{2}}$ is the same as $\sqrt{7}$.

You can similarly show that $7^{\tfrac{1}{3}}$ is the same as $\sqrt[3]{7}$ and that, generally:

$x^{\tfrac{1}{n}}$ is the same as $\sqrt[n]{x}$ (*n*th root of *x*)

So, in summary:

the power $\tfrac{1}{2}$ is the same as the positive square root

the power $\tfrac{1}{3}$ is the same as the cube root

For example:

$49^{\tfrac{1}{2}} = \sqrt{49} = 7$ $\qquad$ $8^{\tfrac{1}{3}} = \sqrt[3]{8} = 2$

$10\,000^{\tfrac{1}{4}} = \sqrt[4]{10\,000} = 10$ $\qquad$ $36^{-\tfrac{1}{2}} = \tfrac{1}{\sqrt{36}} = \tfrac{1}{6}$

EXAMPLE 2

Write $\left(\tfrac{16}{25}\right)^{\tfrac{1}{2}}$ as a fraction.

Find the power of the numerator and denominator separately.

$\left(\tfrac{16}{25}\right)^{\tfrac{1}{2}} = \dfrac{16^{\tfrac{1}{2}}}{25^{\tfrac{1}{2}}} = \dfrac{\sqrt{16}}{\sqrt{25}} = \tfrac{4}{5}$

EXERCISE 1C

TARGETING PASS

1. Evaluate each expression.

a) $25^{\tfrac{1}{2}}$ b) $100^{\tfrac{1}{2}}$ c) $64^{\tfrac{1}{2}}$ d) $81^{\tfrac{1}{2}}$

e) $625^{\tfrac{1}{2}}$ f) $27^{\tfrac{1}{3}}$ g) $64^{\tfrac{1}{3}}$ h) $1000^{\tfrac{1}{3}}$

i) $125^{\tfrac{1}{3}}$ j) $512^{\tfrac{1}{3}}$ k) $144^{\tfrac{1}{2}}$ l) $400^{\tfrac{1}{2}}$

m) $625^{\tfrac{1}{4}}$ n) $81^{\tfrac{1}{4}}$ o) $100000^{\tfrac{1}{5}}$ p) $729^{\tfrac{1}{6}}$

HINTS AND TIPS

You probably know your powers of 2 and know powers of 3, 4 and 5 to 3^3, 4^3, 5^3. If a number is unrecognisable, write out the power tables for 6, 7, 8 etc.

q) $32^{\frac{1}{5}}$ **r)** $1024^{\frac{1}{10}}$ **s)** $1296^{\frac{1}{4}}$ **t)** $216^{\frac{1}{3}}$

u) $16^{-\frac{1}{2}}$ **v)** $8^{-\frac{1}{3}}$ **w)** $81^{-\frac{1}{4}}$ **x)** $3125^{-\frac{1}{5}}$ **y)** $100000^{-\frac{1}{5}}$

TARGETING MERIT

2. Evaluate each expression, giving your answer as a fraction or mixed number.

   a) $\left(\frac{25}{36}\right)^{\frac{1}{2}}$ **b)** $\left(\frac{100}{36}\right)^{\frac{1}{2}}$ **c)** $\left(\frac{64}{81}\right)^{\frac{1}{2}}$ **d)** $\left(\frac{81}{25}\right)^{\frac{1}{2}}$ **e)** $\left(\frac{25}{64}\right)^{\frac{1}{2}}$

   f) $\left(\frac{27}{125}\right)^{\frac{1}{3}}$ **g)** $\left(\frac{8}{512}\right)^{\frac{1}{3}}$ **h)** $\left(\frac{1000}{64}\right)^{\frac{1}{3}}$ **i)** $\left(\frac{64}{125}\right)^{\frac{1}{3}}$ **j)** $\left(\frac{512}{343}\right)^{\frac{1}{3}}$

3. Which of these is the odd one out?

   $16^{-\frac{1}{4}}$ $64^{-\frac{1}{2}}$ $8^{-\frac{1}{3}}$

   Show how you decide.

4. Solve these equations.

   a) $2^x = 8$ **b)** $8^x = 2$ **c)** $4^x = 1$ **d)** $16^x = 4$ **e)** $100^x = 10$
   f) $81^x = 3$ **g)** $16^x = 2$ **h)** $125^x = 5$ **i)** $1000^x = 10$ **j)** $400^x = 20$
   k) $512^x = 8$ **l)** $128^x = 2$

Fractional indices (numerator > 1)

Here are two examples of this form.

$$81^{\frac{3}{4}} = \left(\sqrt[4]{81}\right)^3 = 3^3 = 27 \qquad 16^{\frac{3}{2}} = \left(\sqrt{16}\right)^3 = 4^3 = 64$$

EXAMPLE 3

Evaluate each expression.   **a)** $16^{-\frac{1}{4}}$ **b)** $32^{-\frac{4}{5}}$

When dealing with the negative index remember that it means reciprocal.
Answer problems like these one step at a time.

Step 1: Rewrite the calculation as a fraction by dealing with the negative power.
Step 2: Take the root of the base number given by the denominator of the fraction.
Step 3: Raise the result to the power given by the numerator of the fraction.
Step 4: Write out the answer as a fraction.

a) Step 1: $16^{-\frac{1}{4}} = \left(\frac{1}{16}\right)^{\frac{1}{4}}$ **Step 2:** $16^{\frac{1}{4}} = \sqrt[4]{16} = 2$

   Step 3: $2^1 = 2$ **Step 4:** $16^{-\frac{1}{4}} = \frac{1}{2}$

b) Step 1: $32^{-\frac{4}{5}} = \left(\frac{1}{32}\right)^{\frac{4}{5}}$ **Step 2:** $32^{\frac{1}{5}} = \sqrt[5]{32} = 2$

   Step 3: $2^4 = 16$ **Step 4:** $32^{-\frac{4}{5}} = \frac{1}{16}$

EXAMPLE 4

Write $\left(\frac{8}{27}\right)^{-\frac{2}{3}}$ as a fraction.

$$\left(\frac{8}{27}\right)^{-\frac{2}{3}} = \left(\frac{27}{8}\right)^{\frac{2}{3}} = \frac{27^{\frac{2}{3}}}{8^{\frac{2}{3}}} = \frac{\left(\sqrt[3]{27}\right)^2}{\left(\sqrt[3]{8}\right)^2} = \frac{3^2}{2^2} = \frac{9}{4}$$

The rules for multiplying and dividing with indices still apply for fractional indices. For example:

$$2^{\frac{1}{2}} \times 2^{-2} = 2^{\frac{1}{2} + (-2)} = 2^{-\frac{3}{2}} \qquad 2^{\frac{1}{2}} \div 2^{-2} = 2^{\frac{1}{2} - (-2)} = 2^{\frac{5}{2}}$$

EXERCISE 1D

TARGETING PASS

1. Evaluate each expression.

   a) $32^{\frac{4}{5}}$ b) $125^{\frac{2}{3}}$ c) $1296^{\frac{3}{4}}$ d) $243^{\frac{4}{5}}$

2. Evaluate each expression.

   a) $25^{-\frac{1}{2}}$ b) $36^{-\frac{1}{2}}$ c) $16^{-\frac{1}{4}}$ d) $81^{-\frac{1}{4}}$

   e) $16^{-\frac{1}{2}}$ f) $8^{-\frac{1}{3}}$ g) $32^{-\frac{1}{5}}$ h) $27^{-\frac{1}{3}}$

TARGETING MERIT

3. Evaluate each expression.

   a) $25^{-\frac{3}{2}}$ b) $36^{-\frac{3}{2}}$ c) $16^{-\frac{3}{4}}$ d) $81^{-\frac{3}{4}}$

   e) $64^{-\frac{4}{3}}$ f) $8^{-\frac{2}{3}}$ g) $32^{-\frac{2}{5}}$ h) $27^{-\frac{2}{3}}$

4. Evaluate each expression.

   a) $100^{-\frac{5}{2}}$ b) $144^{-\frac{1}{2}}$ c) $125^{-\frac{2}{3}}$ d) $9^{-\frac{3}{2}}$

   e) $4^{-\frac{5}{2}}$ f) $64^{-\frac{5}{6}}$ g) $27^{-\frac{4}{3}}$ h) $169^{-\frac{1}{2}}$

5. Which of these is the odd one out?

   $16^{-\frac{3}{4}}$ $64^{-\frac{1}{2}}$ $8^{-\frac{2}{3}}$

   Show how you decide.

6. Write each of these as a fraction.

   a) $\left(\frac{9}{4}\right)^{\frac{3}{2}}$ b) $\left(\frac{27}{125}\right)^{\frac{2}{3}}$ c) $\left(\frac{16}{9}\right)^{\frac{5}{2}}$ d) $\left(\frac{4}{49}\right)^{\frac{3}{2}}$

   e) $\left(\frac{64}{27}\right)^{\frac{2}{3}}$ f) $\left(\frac{16}{81}\right)^{\frac{3}{4}}$ g) $\left(\frac{125}{64}\right)^{\frac{4}{3}}$ h) $\left(\frac{64}{729}\right)^{\frac{5}{6}}$

7. Write these as fractions.

a) $\left(\frac{3}{5}\right)^{-2}$
b) $\left(\frac{4}{3}\right)^{-3}$
c) $\left(\frac{9}{5}\right)^{-3}$
d) $\left(\frac{2}{3}\right)^{-5}$
e) $\left(\frac{125}{64}\right)^{-\frac{2}{3}}$
f) $\left(\frac{25}{64}\right)^{-\frac{3}{2}}$
g) $\left(\frac{16}{81}\right)^{-\frac{5}{4}}$
h) $\left(\frac{2187}{128}\right)^{-\frac{5}{7}}$

1.2 Surds

THIS SECTION WILL SHOW YOU HOW TO …

✓ simplify expressions by manipulating surds

KEY WORDS

✓ surd ✓ exact value

A **surd** is a root of a rational number; for example, $\sqrt{5}, \sqrt{6}$. Remember that a rational number is a number that can be written as a fraction where the numerator and denominator are integers.

Surds are also called **exact values**.

Here are some rules of surds.

$$\sqrt{a} \times \sqrt{a} = a \qquad \sqrt{a} \times \sqrt{b} = \sqrt{ab} \qquad \sqrt{a} \div \sqrt{b} = \sqrt{\frac{a}{b}}$$

$$a\sqrt{b} \times c\sqrt{d} = ac\sqrt{bd} \qquad a\sqrt{b} \div c\sqrt{d} = \frac{a}{c}\sqrt{\frac{b}{d}}$$

EXAMPLE 5

Simplify:

a) $\sqrt{6} \times \sqrt{2}$
b) $\sqrt{18} \div \sqrt{2}$

a) $\sqrt{6} \times \sqrt{2} = \sqrt{12}$
$= \sqrt{4 \times 3}$
$= \sqrt{4} \times \sqrt{3}$
$= 2\sqrt{3}$

b) $\sqrt{18} \div \sqrt{2} = \sqrt{\frac{18}{2}} = \sqrt{9} = 3$

HINTS AND TIPS

If possible, factorise the number so that one factor is the square root of a square number. Always write this factor first.

EXAMPLE 6

Simplify:

a) $\sqrt{2} \times \sqrt{3} \times \sqrt{3}$
b) $\sqrt{8} \times \sqrt{3} \div \sqrt{6}$

a) $\sqrt{2} \times \sqrt{3} \times \sqrt{3} = \sqrt{2 \times 3 \times 3} = \sqrt{18} = \sqrt{9}\sqrt{2} = 3\sqrt{2}$

b) $\sqrt{8} \times \sqrt{3} \div \sqrt{6} = \sqrt{8 \times 3 \div 6} = \sqrt{24 \div 6} = \sqrt{4} = 2$

EXERCISE 1E

TARGETING PASS

1. Simplify each expression. Leave your answers in surd form if necessary.
   a) $\sqrt{2} \times \sqrt{7}$ b) $\sqrt{2} \times \sqrt{18}$ c) $\sqrt{6} \times \sqrt{6}$ d) $\sqrt{5} \times \sqrt{6}$

2. Simplify each expression. Leave your answers in surd form if necessary.
   a) $\sqrt{28} \div \sqrt{7}$ b) $\sqrt{48} \div \sqrt{8}$ c) $\sqrt{6} \div \sqrt{6}$ d) $\sqrt{54} \div \sqrt{6}$

3. Simplify each of these surds into the form $a\sqrt{b}$.
   a) $\sqrt{32}$ b) $\sqrt{200}$ c) $\sqrt{1000}$ d) $\sqrt{250}$ e) $\sqrt{98}$ f) $\sqrt{243}$

4. Simplify each expression. Leave your answers in surd form if necessary.
   a) $\sqrt{2} \times \sqrt{3} \times \sqrt{2}$ b) $\sqrt{2} \times \sqrt{2} \times \sqrt{8}$ c) $\sqrt{5} \times \sqrt{8} \times \sqrt{8}$

5. Simplify each expression. Leave your answers in surd form.
   a) $\sqrt{2} \times \sqrt{3} \div \sqrt{2}$ b) $\sqrt{5} \times \sqrt{3} \div \sqrt{15}$ c) $\sqrt{32} \times \sqrt{2} \div \sqrt{8}$
   d) $\sqrt{5} \times \sqrt{8} \div \sqrt{8}$ e) $\sqrt{3} \times \sqrt{3} \div \sqrt{3}$ f) $\sqrt{8} \times \sqrt{12} \div \sqrt{48}$

   HINTS AND TIPS
   Remember to leave your surds in simplified form, e.g. $\sqrt{12} = \sqrt{4}\sqrt{3} = 2\sqrt{3}$

6. Simplify each of these expressions.
   a) $\sqrt{a} \times \sqrt{a}$ b) $\sqrt{a} \div \sqrt{a}$ c) $\sqrt{a} \times \sqrt{a} \div \sqrt{a}$

7. Simplify each of these.
   a) $2\sqrt{18} \times 3\sqrt{2}$ b) $4\sqrt{24} \times 2\sqrt{5}$ c) $3\sqrt{12} \times 3\sqrt{3}$ d) $2\sqrt{8} \times 2\sqrt{8}$
   e) $4\sqrt{2} \times 5\sqrt{3}$ f) $4\sqrt{2} \times 3\sqrt{2}$ g) $2\sqrt{2} \times 2\sqrt{8}$ h) $3\sqrt{3} \times 2\sqrt{3}$

8. Simplify each of these.
   a) $6\sqrt{12} \div 2\sqrt{3}$ b) $3\sqrt{15} \div \sqrt{3}$ c) $6\sqrt{12} \div \sqrt{2}$ d) $4\sqrt{24} \div 2\sqrt{8}$

TARGETING MERIT

9. Simplify each of these.
   a) $4\sqrt{2} \times \sqrt{3} \div 2\sqrt{2}$ b) $4\sqrt{5} \times \sqrt{3} \div \sqrt{15}$ c) $2\sqrt{32} \times 3\sqrt{2} \div 2\sqrt{8}$
   d) $6\sqrt{2} \times 2\sqrt{8} \div 3\sqrt{8}$ e) $3\sqrt{8} \times 3\sqrt{12} \div 3\sqrt{48}$ f) $4\sqrt{7} \times 2\sqrt{3} \div 8\sqrt{3}$

10. Simplify each of these expressions.
    a) $a\sqrt{b} \times c\sqrt{b}$ b) $a\sqrt{b} \div c\sqrt{b}$ c) $a\sqrt{b} \times c\sqrt{b} \div a\sqrt{b}$

11. Find the value of a that makes each of these surds true.
    a) $\sqrt{5} \times \sqrt{a} = 10$ b) $\sqrt{6} \times \sqrt{a} = 12$ c) $\sqrt{10} \times 2\sqrt{a} = 20$
    d) $2\sqrt{6} \times 3\sqrt{a} = 72$

12. Simplify the following.
    a) $\left(\frac{\sqrt{3}}{2}\right)^2$ b) $\left(\frac{5}{\sqrt{3}}\right)^2$ c) $\left(\frac{\sqrt{5}}{4}\right)^2$ d) $\left(\frac{6}{\sqrt{3}}\right)^2$ e) $\left(\frac{\sqrt{8}}{2}\right)^2$

    HINTS AND TIPS
    e.g. $\left(\frac{\sqrt{3}}{2}\right)^2 = \frac{\sqrt{3}}{2} \times \frac{\sqrt{3}}{2}$

13. Decide whether each statement is true or false. Show your working.

   a) $\sqrt{(a+b)} = \sqrt{a} + \sqrt{b}$ **b)** $\sqrt{(a-b)} = \sqrt{a} - \sqrt{b}$

14. Write down a product of two different surds which has an integer answer.

EXAMPLE 7

Write $\sqrt{50} + \sqrt{98}$ in the form $a\sqrt{2}$.

$\sqrt{50} = \sqrt{25 \times 2} = 5\sqrt{2}$ and $\sqrt{98} = \sqrt{49 \times 2} = 7\sqrt{2}$

So $\sqrt{50} + \sqrt{98} = 5\sqrt{2} + 7\sqrt{2} = 12\sqrt{2}$

HINTS AND TIPS

You can only add or subtract surds if they have the same number under the square root sign.

15. a) Write $\sqrt{8} + \sqrt{18}$ in the form $a\sqrt{2}$.

   b) Write $\sqrt{48} - \sqrt{12}$ in the form $b\sqrt{3}$.

   c) Write $\sqrt{50} + \sqrt{72} + \sqrt{18}$ in the form $a\sqrt{2}$.

1.3 Rationalising denominators

THIS SECTION WILL SHOW YOU HOW TO ...

✓ expand brackets which contain surds
✓ rationalise the denominator, including denominators in the form $a\sqrt{b} + c\sqrt{d}$ where a, b, c and d are integers

To rationalise a denominator of the form $a\sqrt{b} + c\sqrt{d}$, multiply the numerator and denominator by $a\sqrt{b} - c\sqrt{d}$.

EXAMPLE 8

a) Rationalise the denominator and simplify $\frac{7}{4\sqrt{2}}$.

b) Rationalise the denominator and simplify $\frac{7}{3-4\sqrt{2}}$.

a) Multiply the numerator and denominator by $\sqrt{2}$ to give:

$$\frac{7}{4\sqrt{2}} \times \frac{\sqrt{2}}{\sqrt{2}} = \frac{7\sqrt{2}}{4\sqrt{4}} = \frac{7\sqrt{2}}{4 \times 2} = \frac{7\sqrt{2}}{8}$$

b) Multiply the numerator and denominator by $3 + 4\sqrt{2}$ to give:

$$\frac{7}{3-4\sqrt{2}} = \frac{7}{(3-4\sqrt{2})} \times \frac{(3+4\sqrt{2})}{3+4\sqrt{2}} = \frac{7(3+4\sqrt{2})}{9-32} = -\frac{7}{23}(3+4\sqrt{2})$$

EXAMPLE 9

$\dfrac{21}{5+x\sqrt{2}}$ simplifies to $15-y\sqrt{2}$ where $x > 0$.

Work out the value of x and the value of y.

Method 1

Multiplying the numerator and denominator by $5-x\sqrt{2}$ gives

$$\dfrac{21}{5+x\sqrt{2}} = \dfrac{21}{(5+x\sqrt{2})} \times \dfrac{(5-x\sqrt{2})}{(5-x\sqrt{2})} = \dfrac{21(5-x\sqrt{2})}{25-2x^2} \Rightarrow \dfrac{21\times 5}{25-2x^2} = 15-y\sqrt{2}$$

So $\dfrac{105}{(25-2x^2)} = 15 \Rightarrow 7 = 25 - 2x^2 \Rightarrow x = 3$ and $y = \dfrac{21\times 3}{25-2\times 3^2} = \dfrac{63}{7} = 9$

So $\dfrac{21(5-x\sqrt{2})}{25-2x^2} = 15 - 9\sqrt{2}$ giving $y = 9$.

Method 2

Alternatively, if $\dfrac{21}{5+x\sqrt{2}} \equiv 15 - y\sqrt{2}$,

$21 \equiv (15-y\sqrt{2})(5+x\sqrt{2}) \Rightarrow 21 \equiv 75 + 15x\sqrt{2} - 5y\sqrt{2} - 2xy$

Equating surds gives $15x - 5y = 0 \Rightarrow y = 3x$.
Equating rational parts gives $21 = 75 - 2xy$.
$21 = 75 - 6x^2 \Rightarrow 6x^2 = 54 \Rightarrow x^2 = 9 \Rightarrow x = 3$
$y = 3x \Rightarrow y = 9$

EXERCISE 1F

TARGETING MERIT

1. Show that:
   a) $(2+\sqrt{3})(1+\sqrt{3}) = 5+3\sqrt{3}$ b) $(1+\sqrt{2})(2+\sqrt{3}) = 2+2\sqrt{2}+\sqrt{3}+\sqrt{6}$
   c) $(4-\sqrt{3})(4+\sqrt{3}) = 13$

2. Expand and simplify where possible.
   a) $\sqrt{3}(2-\sqrt{3})$ b) $\sqrt{2}(3-4\sqrt{2})$ c) $\sqrt{5}(2\sqrt{5}+4)$
   d) $3\sqrt{7}(4-2\sqrt{7})$ e) $3\sqrt{2}(5-2\sqrt{8})$ f) $\sqrt{3}(\sqrt{27}-1)$

3. Expand and simplify where possible.
   a) $(1+\sqrt{3})(3-\sqrt{3})$ b) $(2+\sqrt{5})(3-\sqrt{5})$ c) $(3-2\sqrt{7})(4+3\sqrt{7})$
   d) $(2-3\sqrt{5})(2+3\sqrt{5})$ e) $(2+\sqrt{5})^2$ f) $(1-\sqrt{2})^2$

4. Calculate the area of each of these rectangles, simplifying your answers where possible. (The area of a rectangle with length l and width w is $A = l \times w$.)

HINTS AND TIPS

$(2+\sqrt{5})^2 = (2+\sqrt{5})(2+\sqrt{5})$

a) $(1+\sqrt{3})$ cm by $(2-\sqrt{3})$ cm

b) $(2+\sqrt{10})$ cm by $\sqrt{5}$ cm

c) $2\sqrt{3}$ cm by $(1+\sqrt{27})$ cm

5. Rationalise the denominators of these expressions.

a) $\dfrac{1}{\sqrt{3}}$ b) $\dfrac{1}{2\sqrt{3}}$ c) $\dfrac{3}{\sqrt{3}}$ d) $\dfrac{3\sqrt{2}}{\sqrt{8}}$ e) $\dfrac{\sqrt{7}}{\sqrt{3}}$ f) $\dfrac{2-\sqrt{3}}{\sqrt{3}}$ g) $\dfrac{5+2\sqrt{3}}{\sqrt{3}}$

6. a) Expand and simplify each expression.

     i) $(2+\sqrt{3})(2-\sqrt{3})$ ii) $(1-\sqrt{5})(1+\sqrt{5})$ iii) $(\sqrt{3}-1)(\sqrt{3}+1)$

b) What happens in the answers to part **a**? Why?

7. An engineer uses a formula to work out the number of metres of cable he needs to complete a job. His calculator displays the answer as $10\sqrt{70}$. The button for converting this to a decimal is not working.

He has 80 metres of cable. Without using a calculator, decide whether he has enough cable. Show clearly how you decide.

8. Write $(3+\sqrt{2})^2 - (1-\sqrt{8})^2$ in the form $a+b\sqrt{c}$ where a, b and c are integers.

9. $x^2 - y^2 \equiv (x+y)(x-y)$ is an identity which means it is true for any values of x and y whether they are numeric or algebraic.

Show that it is true for $x = 1+\sqrt{2}$ and $y = 1-\sqrt{8}$.

10. Rationalise the denominator and simplify in each part.

a) $\dfrac{1}{3\sqrt{2}+1}$ b) $\dfrac{5}{3-\sqrt{3}}$ c) $\dfrac{3+\sqrt{2}}{3-\sqrt{2}}$ d) $\dfrac{3\sqrt{5}+1}{\sqrt{5}+2}$ e) $\dfrac{\sqrt{2}+\sqrt{3}}{\sqrt{2}-\sqrt{3}}$

HINTS AND TIPS

$\dfrac{1}{3\sqrt{2}+1} \times \dfrac{3\sqrt{2}-1}{3\sqrt{2}-1}$

TARGETING DISTINCTION

11. Work out the values of x and y such that:

a) $(\sqrt{3}+5)(\sqrt{3}+4) = x + y\sqrt{3}$

b) $(4\sqrt{2})^3 + (6\sqrt{3})^3 = x\sqrt{2} + y\sqrt{3}$

c) $\dfrac{2}{\sqrt{5}+1} = x\sqrt{5}+y$

d) $\dfrac{3}{\sqrt{5}+x} = 3\sqrt{5}+y$.

Exam-style questions

1. Simplify:
   a) $(4\sqrt{5})^2$ [1 mark]
   b) $(\sqrt{8}+1)(1-\sqrt{2})$ [3 marks]

2. a) Write down the value of $64^{\frac{1}{3}}$. [1 mark]
   b) Write down the value of $\left(\frac{1}{64}\right)^{-\frac{1}{3}}$. [1 mark]

3. Given that $81\sqrt{3} = 3^x$, find the value of x, giving your answer as a fraction. [3 marks]

4. Show that $(5+\sqrt{2})(5-\sqrt{2}) = 23$. [2 marks]

5. $x = 2-\sqrt{5}$ and $y = 4-\sqrt{5}$

   Write the following in the form $p+q\sqrt{5}$, where p and q are rational numbers.
   a) xy [3 marks]
   b) $\frac{x}{y}$ [4 marks]

6. Find the value of:
   a) $100^{-\frac{3}{2}}$ [2 marks]
   b) $1000^{\frac{2}{3}}$ [2 marks]
   c) $\left(\frac{27}{8}\right)^{-\frac{2}{3}}$ [3 marks]

7. $a = 1-\sqrt{3}$ and $b = 2+\sqrt{3}$
   Show that $\frac{a+b}{b^2} = 3(7-4\sqrt{3})$. [4 marks]

8. The rectangle has a height of $3-\sqrt{2}$ cm and an area of $4+\sqrt{2}$ cm².

   Area = $4 + \sqrt{2}$ cm² | $3 - \sqrt{2}$ cm

   Show that the perimeter of the rectangle is an integer. [4 marks]

9. Which has the greater value, $\left(\frac{243}{3125}\right)^{-\frac{2}{5}}$ or $\left(\frac{9}{\sqrt{27}}\right)^2$? [5 marks]

2 Algebraic manipulation

2.1 Algebraic indices

THIS SECTION WILL SHOW YOU HOW TO ...
✓ apply the laws of indices
✓ recognise, for example, that x^{-n} is equivalent to $\frac{1}{x^n}$
✓ recognise, for example, that $x^{\frac{a}{b}}$ is equivalent to $(\sqrt[b]{x})^a$

KEY WORDS
✓ index   ✓ indices   ✓ powers   ✓ reciprocal

In Chapter 1, you saw the rules of indices applied to numbers.
The same rules can be applied to algebraic terms.
To *multiply* **powers** of the same variable, *add* the indices.
$a^x \times a^y = a^{(x+y)}$
To *divide* powers of the same variable, *subtract* the indices.
$a^x \div a^y = a^{(x-y)}$
When you *raise* a power to a further power, you *multiply* the indices.
$(a^x)^y = a^{xy}$
Negative **indices** can be written as **reciprocals**. For example:
$x^{-a} = \dfrac{1}{x^a}$

The rules for multiplying and dividing with indices still apply for fractional indices.
For example:
$a^{\frac{1}{2}} \times a^{-2} = a^{-\frac{3}{2}}$ $a^{\frac{1}{2}} \div a^{-2} = a^{\frac{5}{2}}$

Indices of the form $\frac{1}{n}$

The power $\frac{1}{n}$ is the same as the *n*th root.
Here is an example of this form.
$t^{\frac{2}{3}} = t^{\frac{1}{3}} \times t^{\frac{1}{3}} = (\sqrt[3]{t})^2$

Indices of the form $\left(\dfrac{a}{b}\right)^n$

To find the value of an expression in the form $\left(\dfrac{a}{b}\right)^n$, you can work out $\dfrac{a^n}{b^n}$ and then write the answer as a fraction.

To find the value of an expression if the form $\left(\dfrac{a}{b}\right)^{-n}$, invert it and then work out the answer as a fraction:

$$\left(\dfrac{a}{b}\right)^{-n} = \left(\dfrac{b}{a}\right)^{n}$$

EXAMPLE 1

Simplify these expressions. Write your answers in **index** form.

a) $\dfrac{24a^2}{(2a)^3}$ b) $\sqrt[3]{b^5}$ c) $c^{-\frac{1}{2}} \div c$ d) $\dfrac{1}{d^{-3}}$

a) $\dfrac{24a^2}{(2a)^3} = \dfrac{24a^2}{8a^3}$

$= \dfrac{3}{a}$

$= 3\,a^{-1}$

b) $\sqrt[3]{b^5} = b^{\frac{5}{3}}$ c) $c^{-\frac{1}{2}} \div c = c^{\left(-\frac{1}{2} - 1\right)}$ d) $\dfrac{1}{d^{-3}} = d^3$

$= c^{-\frac{3}{2}}$

> **HINTS AND TIPS**
>
> If you move a power from top to bottom, or vice versa, the sign changes. A negative power means the reciprocal: it does not mean that the answer is negative.

EXAMPLE 2

Simplify these expressions. Write your answers in index form.

a) $\dfrac{8x^3 - 12x^5}{4x^2}$ b) $\sqrt{\dfrac{x^2 \times x^{-\frac{1}{2}}}{x^{\frac{1}{2}}}}$ c) $\left(\dfrac{3}{x^{-\frac{2}{3}}}\right)^3 \div \left(9x^3\right)^{\frac{1}{2}}$

a) First, write the fraction as two separate fractions over the same denominator:

$$\dfrac{8x^3 - 12x^5}{4x^2} = \dfrac{8x^3}{4x^2} - \dfrac{12x^5}{4x^2}$$

Then simplify both fractions: $= 2x - 3x^3$

b) First, simplify the numerator by multiplying the terms x^2 and $x^{-\frac{1}{2}}$ (add the indices)

$$\sqrt{\dfrac{x^2 \times x^{-\frac{1}{2}}}{x^{\frac{1}{2}}}} = \sqrt{\dfrac{x^{\frac{3}{2}}}{x^{\frac{1}{2}}}}$$

Next, divide $x^{\frac{3}{2}}$ by $x^{\frac{1}{2}}$ (subtract the indices): $= \sqrt{x}$

Remember that x^1 can be written as x.

Finally, write $\sqrt{x}$ in index form: $= x^{\frac{1}{2}}$

c) First, apply the powers to both terms and simplify:

$$\left(\frac{3}{x^{-\frac{2}{3}}}\right)^3 = \frac{27}{x^{-2}} = 27x^2$$

$$(9x^3)^{\frac{1}{2}} = 3x^{\frac{3}{2}}$$

Now divide: $27x^2 \div 3x^{\frac{3}{2}} = 9x^{\frac{1}{2}}$

EXERCISE 2A

TARGETING PASS

1. Simplify these expressions.
  a) $4c \times 3c^2$
  b) $(2k^3)^2$
  c) $y^{-2} \times 3y$
  d) $4w^{-3} \div 4w^{-1}$
  e) $56d^7 \div 7d^2$
  f) $8d^3 \div (2d^5)^2$
  g) $\frac{18a}{3a^2}$
  h) $\frac{24p^{-5}}{6p^2}$
  i) $\frac{3r^{-5}}{(3r)^2}$

TARGETING MERIT

2. Rewrite the following in index form.
  a) $\sqrt[3]{t^2}$
  b) $\sqrt[4]{m^3}$
  c) $\sqrt[5]{k^2}$
  d) $\sqrt[2]{h^{-3}}$

3. Simplify each expression.
  a) $x^{\frac{3}{2}} \times x^{\frac{5}{2}}$
  b) $x^{\frac{1}{2}} \times x^{-\frac{3}{2}}$
  c) $(8y^3)^{\frac{2}{3}}$
  d) $5x^{\frac{3}{2}} \div \frac{1}{2}x^{-\frac{1}{2}}$
  e) $4x^{\frac{1}{2}} \times 5x^{-\frac{3}{2}}$
  f) $\left(\frac{27}{y^3}\right)^{-\frac{1}{3}}$

4. Simplify each expression.
  a) $x^{\frac{1}{2}} \times x^{\frac{1}{2}}$
  b) $d^{-\frac{1}{2}} \times d^{-\frac{1}{2}}$
  c) $t^{\frac{1}{2}} \times t$
  d) $\left(x^{\frac{1}{2}}\right)^4$
  e) $(y^2)^{\frac{1}{4}}$
  f) $a^{\frac{1}{2}} \times a^{\frac{3}{2}} \times a^2$

TARGETING DISTINCTION

5. Simplify each expression.
  a) $x \div x^{\frac{1}{2}}$
  b) $y^{\frac{1}{2}} \div y^{1\frac{1}{2}}$
  c) $a^{\frac{1}{3}} \times a^{\frac{4}{3}}$
  d) $t^{-\frac{1}{2}} \times t^{-\frac{3}{2}}$
  e) $\frac{1}{d^{-2}}$
  f) $\frac{k^{\frac{1}{2}} \times k^{\frac{3}{2}}}{k^2}$

6. Simplify these expressions.
  a) $8a^3 \times (2a)^{-2}$
  b) $\frac{2}{3}a^{-5} \times 3a^{-2} \times 2a^2$
  c) $15a^5 \div 5a^{\frac{1}{2}} \times \frac{2}{3}a^{-3}$

7. Simplify each expression. Write your answers in index form.

   a) $\dfrac{10x^3 - 12x}{2x^3}$ b) $\dfrac{9x^7 - 6x^2}{3x}$

> **HINTS AND TIPS**
>
> Remember that x can be written as x^1. When simplifying a fraction, remember to divide all the terms in the numerator by the denominator.

TARGETING DISTINCTION *

8. Simplify each expression.

   a) $\sqrt{x^{\frac{3}{2}} \times x^{\frac{9}{2}}}$ b) $\sqrt{x^{\frac{1}{2}} \times x^{-\frac{5}{2}}}$ c) $\sqrt{x^{\frac{4}{3}} \times x^{\frac{2}{3}}}$

   d) $\sqrt{\dfrac{x^{\frac{5}{2}}}{x^{\frac{1}{2}}}}$ e) $\sqrt{\dfrac{x^{-\frac{1}{2}} \times x^5}{x^{\frac{3}{2}}}}$ f) $\sqrt{\dfrac{x^{\frac{1}{3}} \times x^{\frac{7}{3}}}{x^{\frac{2}{3}}}}$

9. Simplify the following.

   a) $8x^{-3} \times \tfrac{1}{2}x^{-2} \div \tfrac{3}{4}x^{-6}$ b) $\left(3x^{\frac{1}{3}}\right)^{-2} \div \dfrac{1}{3x}$ c) $\left(\dfrac{2}{x^{-\frac{1}{2}}}\right)^4 \div \left(4x^3\right)^{\frac{1}{2}}$

2.2 Expanding brackets

THIS SECTION WILL SHOW YOU HOW TO ...
- ✓ expand and simplify two or three brackets
- ✓ expand brackets of the form $(a + b)^n$
- ✓ use Pascal's triangle to expand binomial expressions

KEY WORDS
✓ expand and simplify   ✓ expression   ✓ like terms   ✓ Pascal's triangle

When you **expand and simplify** a product of brackets you should collect **like terms**. Algebraic **expressions** should always be simplified as much as possible.

The rule for expanding expressions such as $(t + 5)(t - 4)(2t - 1)$ is similar to that for expanding single and double brackets: multiply everything in one set of brackets by everything in a second set of brackets, simplify, and then multiply each resultant term by each term in the third bracket (see Example 5).

EXAMPLE 3

Expand and simplify the following.

a) $(x - 3)(x + 7)$ b) $(3x - 5)^2$ c) $(x + 3)(x - 3)$

a) $(x - 3)(x + 7) = x^2 + 7x - 3x - 21$
$\qquad\qquad\qquad\;\, = x^2 + 4x - 21$

Always simplify your answer by gathering together like terms.

b) Remember that when you square an expression you multiply it by itself:
$$(3x - 5)^2 = (3x - 5)(3x - 5)$$
$$= 9x^2 - 15x - 15x + 25$$
$$= 9x^2 - 30x + 25$$

c) $(x + 3)(x - 3) = x^2 + 3x - 3x - 9$
$$= x^2 - 9$$

EXAMPLE 4

Expand and simplify $(x^4 + 2x^3 - 4x^2 + 5x - 1)(x + 1)$.

Multiply every term in the first bracket by every term in the second bracket:
$$= x^4(x + 1) + 2x^3(x + 1) - 4x^2(x + 1) + 5x(x + 1) - 1(x + 1)$$

You should end up with 10 terms ((first bracket 5 terms) × (second bracket 2 terms)).
$$= x^5 + 2x^4 - 4x^3 + 5x^2 - x + x^4 + 2x^3 - 4x^2 + 5x - 1$$

Finally, simplify and write your answer in descending powers of *x*:
$$= x^5 + 3x^4 - 2x^3 + x^2 + 4x - 1$$

EXAMPLE 5

Expand and simplify:

a) $(t + 5)(t - 4)(2t - 1)$ **b)** $(3f + 5)^3$

a) Start by multiplying two brackets: $(t + 5)(t - 4) = (t^2 + t - 20)$

Then multiply the result by the third bracket:
$$= (t^2 + t - 20)(2t - 1)$$
$$= t^2(2t - 1) + t(2t - 1) - 20(2t - 1)$$
$$= 2t^3 - t^2 + 2t^2 - t - 40t + 20$$

Simplify:
$$= 2t^3 + t^2 - 41t + 20$$

b) Applying the power 3 means that you are cubing the expression, so multiply it by itself, simplify and then multiply this answer by itself again.
$$(3f + 5)^3 = (3f + 5)(3f + 5)(3f + 5)$$
$$= (9f^2 + 30f + 25)(3f + 5)$$
$$= 9f^2(3f + 5) + 30f(3f + 5) + 25(3f + 5)$$
$$= 27f^3 + 45f^2 + 90f^2 + 150f + 75f + 125$$
$$= 27f^3 + 135f^2 + 225f + 125$$

EXERCISE 2B

TARGETING PASS

1. Without expanding the brackets, match each expression on the left with the correct expression on the right.

   $(3x - 2)(2x + 1)$ $4x^2 - 4x + 1$

   $(2x - 1)(2x - 1)$ $6x^2 - x - 2$

   $(6x - 3)(x + 1)$ $6x^2 + 7x + 2$

   $(4x + 1)(x - 1)$ $6x^2 + 3x - 3$

   $(3x + 2)(2x + 1)$ $4x^2 - 3x - 1$

2. Try to spot the pattern in each of these expressions so that you can immediately write down the expansion.

   a) $(2x + 1)(2x - 1)$ b) $(5y + 3)(5y - 3)$ c) $(4h - 1)(4h + 1)$

   d) $(2 + 3x)(2 - 3x)$ e) $(6 - 5y)(6 + 5y)$ f) $(a + b)(a - b)$

   g) $(2m - 3p)(2m + 3p)$ h) $(ab + cd)(ab - cd)$ i) $(a^2 + b^2)(a^2 - b^2)$

3. Show that the areas of the shaded regions are the same.

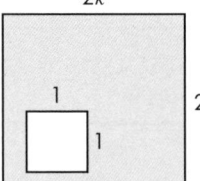

4. Expand and simplify:

   a) $(x^2 + 3x - 1)(x + 1)$
   b) $(2x^2 + 5x + 1)(4x + 1)$
   c) $(x^3 - 6x^2 + 2x - 3)(2x + 3)$
   d) $x^2(x^4 + x^3 - 6x^2 + 2x - 3) + x(2x^2 + 3x - 1)$

TARGETING MERIT

5. Expand and simplify:

   a) $(x + 1)^3$ b) $(2x - 1)^3$ c) $(3x + 2)^3$ d) $(4x - 3)^3$

TARGETING DISTINCTION

6. Expand and simplify:

   a) $2x^{\frac{1}{2}}\left(x^{\frac{1}{2}} - x^{-\frac{1}{2}}\right)$
   b) $x^{\frac{1}{3}}\left(x^{\frac{2}{3}} - x^{-\frac{1}{3}}\right)$
   c) $\left(3x^{\frac{1}{2}} - 2x^{-\frac{1}{2}}\right)\left(2x^{\frac{1}{2}} - 3x^{-\frac{1}{2}}\right)$
   d) $\left(4x^{\frac{1}{2}} - x^{-\frac{1}{3}}\right)\left(4x^{\frac{1}{2}} + x^{-\frac{1}{3}}\right)$.

Here are the first seven rows of **Pascal's triangle**.

Here are some expansions.

$(a + b)^2 = a^2 + 2ab + b^2$

$(a + b)^3 = a^3 + 3a^2b + 3ab^2 + b^3$

$(a + b)^4 = a^4 + 4a^3b + 6a^2b^2 + 4ab^3 + b^4$

$(a + b)^5 = a^5 + 5a^4b + 10a^3b^2 + 10a^2b^3 + 5ab^4 + b^5$

$(a + b)^6 = a^6 + 6a^5b + 15a^4b^2 + 20a^3b^3 + 15a^2b^4 + 6ab^5 + b^6$

```
              1
            1   1
          1   2   1
        1   3   3   1
      1   4   6   4   1
    1   5  10  10   5   1
  1   6  15  20  15   6   1
```

Compare the coefficients with the numbers in Pascal's triangle.

Pascal's triangle is a triangular pattern of coefficients which correspond to the coefficients in binomial expansions.

It can be used to save time when expanding brackets and to find specific coefficients in expansions.

EXAMPLE 6

Use Pascal's triangle to expand $(2x + 5)^4$.

Substituting $a = 2x$ and $b = 5$ into

$(a + b)^4 = \mathbf{1}a^4 + \mathbf{4}a^3b + \mathbf{6}a^2b^2 + \mathbf{4}ab^3 + \mathbf{1}b^4$

gives

$(2x + 5)^4 = \mathbf{1}(2x)^4 + \mathbf{4}(2x)^3(5) + \mathbf{6}(2x)^2(5)^2 + \mathbf{4}(2x)(5)^3 + \mathbf{1}(5)^4$

$= 16x^4 + 160x^3 + 600x^2 + 1000x + 625$

Notice that the powers of a descend from 4 to 0 and the powers of b ascend from 0 to 4.

The sum of the indices of a and b in each term equal 4.

EXAMPLE 7

Work out the coefficient of x^3 in the expansion of $(3x - 4)^5$.

Substitute $a = 3x$ and $b = -4$ into

$(a + b)^5 = \mathbf{1}a^5 + \mathbf{5}a^4b + \mathbf{10}a^3b^2 + \mathbf{10}a^2b^3 + \mathbf{5}ab^4 + \mathbf{1}b^5$

You are only interested in the term x^3 so you can just look at $\mathbf{10}a^3b^2$ which will contain x^3

$\mathbf{10}a^3b^2 = 10(3x)^3(-4)^2$

$= 10 \times 27x^3 \times 16$

$= 4320x^3$

The coefficient of x^3 in the expansion of $(3x - 4)^5$ is 4320.

EXERCISE 2C

TARGETING DISTINCTION*

1. Use Pascal's triangle to expand $(x - 3)^4$.
2. Work out the coefficient of x^3 in the expansion of $(2 + 5x)^5$.
3. Work out the coefficient of x^5 in the expansion of $(4x - 1)^6$.
4. Work out the coefficient of x^5 in the expansion of $(3x + 2)^6$.
5. The coefficient of x^3 in the expansion of $(3x + a)^5$ is 1080. Work out the value of a.
6. In the expansion of $(2x + a)^4$ the coefficient of x^2 is equal to the coefficient of x. Work out the value of a.
7. Use Pascal's triangle to expand $(1 + \frac{1}{2}x)^4$.
8. The coefficient of x^5 in the expansion of $(2x + p)^6$ is 960. Find the value of p.

HINTS AND TIPS

Write out the line you are using from Pascal's triangle to help you; don't forget to use brackets in your expressions when substituting.

2.3 Factorising

THIS SECTION WILL SHOW YOU HOW TO ...

✓ factorise quadratic expressions
✓ understand and use the factor theorem to factorise polynomials
✓ show that $ax - b$ is a factor of the function f(x) by checking that $f\left(\frac{b}{a}\right) = 0$

KEY WORDS

✓ brackets
✓ factorisation
✓ cubic
✓ coefficient
✓ quadratic expression
✓ degree
✓ difference of two squares
✓ polynomial

Quadratic factorisation

Factorisation involves changing a **quadratic expression** into the product of two linear expressions, written as two terms in **brackets**. Start with the factorisation of quadratic expressions of type $x^2 + ax + b$, where a and b are integers.

Here are some simple rules that will help you to factorise.

- The expression inside each set of brackets will start with an x, and the signs in the quadratic expression show which signs to put after the x.
- When the second sign in the expression is a plus, the signs in both sets of brackets are the same as the first sign.

$x^2 + ax + b = (x + ?)(x + ?)$ since everything is positive

$x^2 - ax + b = (x - ?)(x - ?)$ since −ve × −ve = +

- When the second sign is a minus, the signs in the brackets are different.
  $x^2 + ax - b = (x + ?)(x - ?)$ since +ve × −ve = −ve
  $x^2 - ax - b = (x + ?)(x - ?)$
- Next, look at the last number, b, in the expression. When multiplied together, the two numbers in the brackets must give b.
- Finally, look at the **coefficient** of x, which is a. The sum of the two numbers in the brackets will give a.

EXAMPLE 8

Factorise:

a) $x^2 - x - 6$ **b)** $x^2 - 9x + 20$

a) Because of the signs, you know that $x^2 - x - 6 = (x + ?)(x - ?)$.

$a = -1$ and $b = -6$, so you need two numbers that have a product of -6 and a sum of -1. These are $+2$ and -3.

So, $x^2 - x - 6 = (x + 2)(x - 3)$.

b) Because of the signs, you know that $x^2 - 9x + 20 = (x - ?)(x - ?)$.

$a = -9$ and $b = +20$, so you need two numbers that have a product of $+20$ and a sum of -9. These are -4 and -5.

So, $x^2 - 9x + 20 = (x - 4)(x - 5)$.

Difference of two squares

$a^2 - b^2 = (a + b)(a - b)$

This type of quadratic expression, with only two terms, both of which are perfect squares separated by a minus sign, is called the **difference of two squares**. For example:

$x^2 - 9$ $x^2 - 25$ $x^2 - 4$ $x^2 - 100$

There are three conditions that must be met for an expression to be the difference of two squares.

- There must be two terms.
- They must be separated by a negative sign.
- Each term must be a perfect square, for example, x^2 and $9n^2$.

EXAMPLE 9

Factorise $9x^2 - 169$.

This expression is in the form $a^2 - b^2$ and so can be factorised using the difference of two squares

$$a^2 - b^2 = (a + b)(a - b)$$

The two squares in the expression are $(3x)^2$ and 13^2.

So, it factorises to $(3x + 13)(3x - 13)$.

EXERCISE 2D

TARGETING PASS

Factorise the expressions in questions **1–12**.

1. $y^2 + 5y - 6$
2. $m^2 - 4m - 12$
3. $n^2 - 3n - 18$
4. $m^2 - 7m - 44$
5. $t^2 - t - 90$
6. $h^2 - h - 72$
7. $t^2 - 2t - 63$
8. $y^2 + 20y + 100$
9. $m^2 - 18m + 81$
10. $x^2 - 24x + 144$
11. $d^2 - d - 12$
12. $q^2 - q - 56$

TARGETING MERIT

Each of the expressions in questions **13–18** is the difference of two squares. Factorise each one.

13. $x^2 - 4y^2$
14. $9x^2 - 1$
15. $16x^2 - 9$
16. $25x^2 - 64y^4$
17. $\frac{1}{4}x^2 - \frac{1}{9}y^2$
18. $\frac{4}{81}x^2 - 100$

Factorising $ax^2 + bx + c$

You can factorise quadratic expressions in the form $ax^2 + bx + c$ ($a \neq 1$) using the grouping method:

- Find two factors of ac that add up to b.
- Write the term b as the sum of these two factors.
- Group the first and second term together and group the third and fourth term together.
- Factorise each group of two terms.
- Write as a pair of brackets.

EXAMPLE 10

Factorise $3x^2 + 8x + 4$.

$a = 3$, $b = 8$, $c = 4$ and $ac = 12$

Two factors of ac that sum to b are 6 and 2: ($6 \times 2 = 12$ and $6 + 2 = 8$)

$bx = 8x$ and can be written as $6x + 2x$, so the expression becomes $3x^2 + 6x + 2x + 4$.

Grouping the first and second terms and the third and fourth terms together gives $(3x^2 + 6x) + (2x + 4)$.

Factorising each group gives $\mathbf{3x}(x + 2) + \mathbf{2}(x + 2)$. (The expressions in the brackets must be the same.)

Therefore,

$3x^2 + 8x + 4 = (\mathbf{3x + 2})(x + 2)$

You can check this result by multiplying out the brackets: $(3x + 2)(x + 2) = 3x^2 + 8x + 4$

EXAMPLE 11

Factorise $6x^2 - 7x - 10$.

$a = 6$, $b = -7$, $c = -10$ and $ac = -60$

Two factors of ac that add up to b are 5 and -12.

$bx = -7x$ can be written as $5x - 12x$, so the expression can be written as $6x^2 + 5x - 12x - 10$.

Grouping the first and second terms and the third and fourth terms together gives $(6x^2 + 5x) - (12x + 10)$.

Factorising each group gives $x(6x + 5) - 2(6x + 5)$.

Therefore,

$6x^2 - 7x - 10 = (x - 2)(6x + 5)$

You can check this result by multiplying out the brackets: $(x - 2)(6x + 5) = 6x^2 - 7x - 10$

EXERCISE 2E

TARGETING MERIT

Factorise the expressions in questions **1–12**.

1. $2x^2 + 5x + 2$
2. $7x^2 + 8x + 1$
3. $4x^2 + 3x - 7$
4. $24t^2 + 19t + 2$
5. $15t^2 + 2t - 1$
6. $16x^2 - 8x + 1$
7. $6y^2 + 33y - 63$
8. $4y^2 + 8y - 96$
9. $8x^2 + 10x - 3$
10. $6t^2 + 13t + 5$
11. $3x^2 - 16x - 12$
12. $-7x^2 + 37x - 10$

13. This rectangle is made up of four parts, with areas of $12x^2$, $3x$, $8x$ and 2 square units.

    Work out expressions for the sides of the rectangle, in terms of x.

    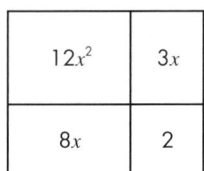

14. Three students are asked to factorise the expression $6x^2 + 30x + 36$. These are their answers.

Adam	**Bertie**	**Cara**
$(6x + 12)(x + 3)$	$(3x + 6)(2x + 6)$	$(2x + 4)(3x + 9)$

    All the answers are correctly factorised.

    a) Explain why one quadratic expression can have three different factorisations.

    b) Which of the following is the most complete factorisation?

    $2(3x + 6)(x + 3)$ $6(x + 2)(x + 3)$ $3(x + 2)(2x + 6)$

    Explain your choice.

EXAMPLE 12

Factorise $3x^2 + 8xy + 4y^2$.

$3x^2 + 8x + 4 = (3x + 2)(x + 2)$

So $3x^2 + 8xy + 4y^2 = (3x + 2y)(x + 2y)$

You can check this by expanding $(3x + 2y)(x + 2y)$.

HINTS AND TIPS

This is similar to Example 10. Look at the pattern with the y terms.

EXAMPLE 13

Factorise $6x^2 - 7xy - 10y^2$.

$$6x^2 - 7x - 10 = (6x + 5)(x - 2)$$

So $6x^2 - 7xy - 10y^2 = (6x + 5y)(x - 2y)$

You can check this by expanding $(6x + 5y)(x - 2y)$.

HINTS AND TIPS

This is similar to Example 11. Look at the pattern with the y terms.

EXAMPLE 14

Factorise fully $(3x + 5)^2 - (2x - 3)^2$.

This is a difference of two squares, $(3x + 5)^2$ and $(2x - 3)^2$.

$(3x + 5)^2 - (2x - 3)^2 = (3x + 5 + 2x - 3)(3x + 5 - (2x - 3))$
$= (5x + 2)(x + 8)$

EXAMPLE 15

Factorise $2p^4 + \frac{5}{4}p^2q - \frac{3}{4}q^2$.

It is helpful here to use $x = p^2$ giving $2x^2 + \frac{5}{4}xq - \frac{3}{4}q^2$

To make the factorising easier, you can take out the common factor of $\frac{1}{4}$ to leave integers:

$$2x^2 + \frac{5}{4}xq - \frac{3}{4}q^2 = \frac{1}{4}(8x^2 + 5qx - 3q^2)$$

You can now factorise this using the grouping method.

$a = 8$, $b = 5q$, $c = -3q^2$ and $ac = -24q^2$

Two factors of ac that sum to b are $8q$ and $-3q$ ($8q \times -3q = -24q^2$, and $8q + -3q = 5q$)

$bx = 5qx$ and can be written as $8qx - 3qx$ giving $\frac{1}{4}(8x^2 + 8qx - 3qx - 3q^2)$

Grouping the first two terms and the second two terms together gives
$\frac{1}{4}((8x^2 + 8qx) - (3qx + 3q^2))$

Factorising each group gives $\frac{1}{4}(8x(x + q) - 3q(x + q))$

Therefore,

$$2x^2 + \frac{5}{4}xq - \frac{3}{4}q^2 = \frac{1}{4}(8x - 3q)(x + q)$$

Finally, replace x with p^2 giving

$\frac{1}{4}(8p^2 - 3q)(p^2 + q)$, which can be written as $(2p^2 - \frac{3}{4}q)(p^2 + q)$

EXERCISE 2F

TARGETING DISTINCTION

Factorise the expressions in questions **1–9**.

1. $15x^2 + 2xy - y^2$
2. $x^4 - 25y^4$
3. $8x^3 - 50x$
4. $16x^4 - 25y^4$
5. $(x + 5)^2 - (x - 3)^2$
6. $(2x + 1)^2 - (2x - 1)^2$
7. $(3x + 2)^2 - (2x + 3)^2$
8. $4x^2 - (2x + 1)^2$
9. $(5x + 1)^2 - 9x^2$

TARGETING DISTINCTION*

Factorise the expressions in questions **10–12**.

10. $14x^4 - 17x^2y - 6y^2$
11. $6x^5 - 14x^4y + 4x^3y^2$
12. $\frac{1}{4}p^4 + \frac{3}{2}p^2q + 2q^2$

HINTS AND TIPS

To factorise expressions with different indices (e.g. x^4 and y^2) replace with familiar terms (e.g. a^2 and b), then factorise and substitute back when completed.

The factor theorem

A **polynomial** is an expression with terms that have positive integer powers of the variable and possibly a constant term.

Here are some examples of polynomials.

- $x^2 - 2x - 3$ This is also called a quadratic, as the highest power is 2.
- $x^3 - 3x^2 + 2x$ This is also called a **cubic**, as the highest power is 3.
- $x^5 + 4x^4 - x^3 + 3x^2 + 1$ This is called a **polynomial of degree 5**, as the highest power is 5.

The quadratic expression $x^2 - 2x - 3$ can be factorised to give $(x - 3)(x + 1)$.

The solutions to the quadratic equation $x^2 - 2x - 3 = 0$ can be found by factorising the left-hand side.

This gives $(x - 3)(x + 1) = 0$

So either $x - 3 = 0$ or $x + 1 = 0$

Therefore, the solutions are $x = 3$ and $x = -1$.

Now suppose that the solutions of a quadratic equation are $x = -4$ and $x = 5$.

By reversing the process shown above, this gives: $x + 4 = 0$ or $x - 5 = 0$

This leads to $(x + 4)(x - 5) = 0$ and expanding the brackets and simplifying gives $x^2 - x - 20 = 0$.

So, the quadratic equation with solutions $x = -4$ and $x = 5$ is $x^2 - x - 20 = 0$.

You can write the expression on the left-hand side using function notation as $f(x) = x^2 - x - 20$.

Because $x = -4$ and $x = 5$ are solutions to the quadratic equation $x^2 - x - 20 = 0$, $f(-4) = (-4)^2 - (-4) - 20 = 0$ and $f(5) = 5^2 - 5 - 20 = 0$.

This can be extended to cubics and other polynomials.

In general, if f(x) is a polynomial and a value a can be found so that f(a) = 0, then $x = a$ is a solution of the equation f(x) = 0 and (x − a) is a factor of the polynomial.
Also, if (ax − b) is a factor of the function f(x) then $f\left(\frac{b}{a}\right) = 0$.

EXAMPLE 16

Factorise $x^3 - 6x^2 + 11x - 6$.
Let f(x) = $x^3 - 6x^2 + 11x - 6$.
f(1) = $1^3 - 6(1)^2 + 11(1) - 6 = 1 - 6 + 11 - 6 = 0$
So, as f(1) = 0, (x − 1) is a factor of $x^3 - 6x^2 + 11x - 6$.
The constant term is −6 so try the other factors of −6: these are 2, 3, −1, −2, −3 and −6.
f(2) = $2^3 - 6(2)^2 + 11(2) - 6 = 8 - 24 + 22 - 6 = 0$
So, as f(2) = 0, (x − 2) is a factor of $x^3 - 6x^2 + 11x - 6$.
f(3) = $3^3 - 6(3)^2 + 11(3) - 6 = 27 - 54 + 33 - 6 = 0$
So, as f(3) = 0, (x − 3) is a factor of $x^3 - 6x^2 + 11x - 6$.
Therefore, $x^3 - 6x^2 + 11x - 6 = (x - 1)(x - 2)(x - 3)$.

EXAMPLE 17

Show that (x − 4) is a factor of $x^4 - x^3 - 10x^2 - 7x - 4$.
Let f(x) = $x^4 - x^3 - 10x^2 - 7x - 4$.

If (x − 4) is a factor, then $x - 4 = 0 \Rightarrow x = 4$.
Substitute x = 4 into f(x).

f(4) = $(4)^4 - (4)^3 - 10(4)^2 - 7(4) - 4 = 256 - 64 - 160 - 28 - 4 = 0$
So, as f(4) = 0, (x − 4) is a factor of $x^4 - x^3 - 10x^2 - 7x - 4$.

EXAMPLE 18

Show that 2x − 1 is a factor of $2x^3 - 11x^2 + 13x - 4$.
Let f(x) = $2x^3 - 11x^2 + 13x - 4$.

If (2x − 1) is a factor, then $2x - 1 = 0 \Rightarrow x = \frac{1}{2}$
Substitute $x = \frac{1}{2}$ into f(x)

$f\left(\frac{1}{2}\right) = 2\left(\frac{1}{2}\right)^3 - 11\left(\frac{1}{2}\right)^2 + 13\left(\frac{1}{2}\right) - 4 = \frac{1}{4} - \frac{11}{4} + \frac{13}{2} - 4 = 0$

So, as $f\left(\frac{1}{2}\right) = 0$, (2x − 1) is a factor of $2x^3 - 11x^2 + 13x - 4$.

EXERCISE 2G

TARGETING DISTINCTION*

1. In each part, show that the expression is a factor of the polynomial.
   a) $x + 1, x^3 + 6x^2 - 9x - 14$
   b) $x - 3, x^3 + 3x^2 - 13x - 15$
   c) $x - 4, x^3 - 7x^2 + 2x + 40$
   d) $x + 6, x^3 + 13x^2 + 54x + 72$
   e) $x + 7, x^3 - 37x + 84$
   f) $2x - 3, 2x^3 - 5x^2 + x + 3$

2. Factorise each expression.
   a) $x^3 + 3x^2 - 13x - 15$
   b) $x^3 + 5x^2 + 2x - 8$
   c) $x^3 + 6x^2 + 11x + 6$
   d) $x^3 + 6x^2 - x - 30$
   e) $2x^3 - 13x^2 + 17x + 12$
   f) $6x^3 + 19x^2 + 11x - 6$

EXAMPLE 19

The expression $x^3 - 4x^2 + ax + b$ has the factors $(x - 2)$ and $(x + 3)$.
Work out the third factor of the expression.

Because $(x - 2)$ is a factor of f(x), using the factor theorem, f(2) = 0
Substituting this value for x gives $\quad 2^3 - 4(2)^2 + 2a + b = 0 \quad$ so $\quad 2a + b = 8$
Also, because $(x + 3)$ is a factor of f(x), using the factor theorem, f(−3) = 0.
Substituting this value for x gives $\quad (-3)^3 - 4(-3)^2 - 3a + b = 0 \quad$ so $\quad -3a + b = 63$
Now solve these two equations simultaneously:

$$2a + b = 8 \quad (1)$$
$$-3a + b = 63 \quad (2)$$

Subtract (1) − (2): $\quad\quad\quad\quad\quad\quad 5a = -55$
Solve for a: $\quad\quad\quad\quad\quad\quad\quad a = -11$
Substituting the value of a into (1): $\quad 2(-11) + b = 8$
$\quad\quad\quad\quad\quad\quad\quad\quad\quad\quad\quad\quad b = 30$

There are now two methods of finding the third factor of f(x).
You can use algebraic division to divide f(x) by $(x - 2)$ and $(x + 3)$.
Alternatively, you can compare coefficients.
Since $a = -11$ and $b = 30$,
$$x^3 - 4 - 11x + 30 = (x - 2)(x + 3)(x + c)$$
The third factor must be in the form $(x + c)$ as the coefficient of x^3 is 1.
The constant of f(x) is 30, which must equal $-2 \times 3 \times c$
Therefore, $-6c = 30$
So $\quad\quad c = -5$
The third factor of f(x) is $(x - 5)$.

TARGETING DISTINCTION*

3. The expression $x^3 - 3x^2 + ax + b$ has the factors $(x - 1)$ and $(x - 4)$.
   Work out the third factor of the expression.

4. a) $x + 5$ is a factor of $x^3 + 3x^2 - 13x + c$. Show that $c = -15$.

   b) Work out the other two factors of the cubic.

5. $x^3 + 3x^2 + ax + b$ factorises as $(x + c)^3$.
   Work the values of a, b and c.

6. $x^3 + 3x^2 - 16x - 48$ factorises as $(x^2 - a^2)(x + b)$.
   Work out the three linear factors of $x^3 + 3x^2 - 16x - 48$.

> **HINTS AND TIPS**
>
> Remember to use Pascal's triangle to work out the coefficients.

7. Work out the common factors of $x^3 - 5x^2 - 2x + 24$ and $x^3 - x^2 - 10x - 8$.

8. Show that $x + 3$ is a factor of $x^4 - 13x^2 + 36$.

9. Show that $2x - 5$ is a factor of $2x^4 + x^3 - 17x^2 - x + 15$.

10. a) Use the factor theorem to show that $3x - 1$ is a factor of $3x^3 - 22x^2 + 43x - 12$.

    b) Hence factorise fully $3x^3 - 22x^2 + 43x - 12$.

2.4 Completing the square

THIS SECTION WILL SHOW YOU HOW TO ...
✓ complete the square

KEY WORDS
✓ completing the square

Quadratic equations can be re-written by **completing the square** using a process involving perfect squares.

Remember that $(x + a)^2 = x^2 + 2ax + a^2$

which can be rearranged to give $x^2 + 2ax = (x + a)^2 - a^2$

This is the principle behind completing the square.

There are three basic steps in rewriting $x^2 + px + q$ in the form $(x + a)^2 + b$.

Step 1: Ignore q and just look at the first two terms, $x^2 + px$.

Step 2: Rewrite $x^2 + px$ as $\left(x + \frac{p}{2}\right)^2 - \left(\frac{p}{2}\right)^2$.

Step 3: Bring back q and simplify the constant to get $x^2 + px + q = \left(x + \frac{p}{2}\right)^2 - \left(\frac{p}{2}\right)^2 + q$.

EXAMPLE 20

Rewrite the following in the form $(x \pm a)^2 \pm b$.
a) $x^2 + 6x - 7$ b) $x^2 - 8x + 3$

a) Ignore -7 for the moment and just consider $x^2 + 6x$.

Rewrite this as $x^2 + 6x = (x + 3)^2 - 9$ (You need to subtract 9 to remove the constant term.)

You can check that this is equivalent to the original expression:
$$(x + 3)^2 - 9 = x^2 + 6x + 9 - 9$$
$$= x^2 + 6x$$

Now bring back the -7, so $x^2 + 6x - 7 = (x + 3)^2 - 9 - 7$

Combine the constant terms to get the final answer:
$$x^2 + 6x - 7 = (x + 3)^2 - 16$$

b) Ignore $+3$ for the moment and just consider $x^2 - 8x$.

Rewrite this as $x^2 - 8x = (x - 4)^2 - 16$ (You still subtract $(-4)^2$, as $(-4)^2 = +16$)

Now bring the $+3$ back, so $x^2 - 8x + 3 = (x - 4)^2 - 16 + 3$

Combine the constant terms to get the final answer:
$$x^2 - 8x + 3 = (x - 4)^2 - 13$$

HINTS AND TIPS

When the coefficient of the x^2 term is not 1, so that the expression is of the form $ax^2 + bx + c$, start by taking a factor of a outside a bracket.

EXAMPLE 21

Rewrite $3x^2 + 12x - 6$ in the form $a(x + b)^2 + c$.

Start by ignoring the '-6' term.

Factorise $3x^2 + 12x$: $3x^2 + 12x = 3(x^2 + 4x)$

Rewrite $x^2 + 4x$ as $(x + 2)^2 - 4$: $= 3((x + 2)^2 - 4)$

Expand the outer brackets: $= 3(x + 2)^2 - 12$

Now bring back the '-6' term: $3x^2 + 12x - 6 = 3(x + 2)^2 - 12 - 6$
$$= 3(x + 2)^2 - 18$$

EXERCISE 2H

TARGETING MERIT

1. Write an equivalent expression in the form $(x \pm a)^2 - b$.
   a) $x^2 + 4x$
   b) $x^2 + 14x$
   c) $x^2 - 6x$
   d) $x^2 + 6x$
   e) $x^2 - 3x$
   f) $x^2 - 9x$
   g) $x^2 + 13x$
   h) $x^2 + 10x$
   i) $x^2 + 8x$
   j) $x^2 - 2x$
   k) $x^2 + 2x$
   l) $x^2 - 5x$

2. Write an equivalent expression in the form $(x \pm a)^2 - b$.
   Question **1** will help.
   a) $x^2 + 4x - 1$
   b) $x^2 + 14x - 5$
   c) $x^2 - 6x + 3$
   d) $x^2 + 6x + 7$
   e) $x^2 - 3x - 1$
   f) $x^2 - 9x + 10$
   g) $x^2 + 13x + 35$
   h) $x^2 + 10x + 3$
   i) $x^2 + 8x - 6$
   j) $x^2 + 2x - 1$
   k) $x^2 - 2x - 7$
   l) $x^2 + 2x - 9$

TARGETING DISTINCTION

3. Rewrite each equation in the form $a(x + b)^2 + c$.
   a) $2x^2 + 4x + 7$
   b) $3x^2 + 12x + 3$
   c) $6x^2 + 12x + 4$
   d) $5x^2 - 30x + 12$
   e) $8x^2 - 32x + 10$
   f) $9x^2 + 9x + 9$
   g) $12x^2 - 36x + 14$
   h) $5x^2 + 10x + 6$
   i) $7x^2 + 14x + 5$
   j) $7x^2 + 7x + 2$
   k) $10x^2 - 20x + 5$
   l) $11x^2 + 22x + 6$

4. Work out the values of a, b and c such that $ax^2 + bx + 10 \equiv (2x + 1)^2 + c$.

5. Work out the values of a, b and c such that $4x^2 + ax - 5 \equiv b(x + 3)^2 + c$.

6. Work out the values of a and b such that $5 - 4x - x^2 \equiv a - (x + b)^2$.

7. Ahmed rewrites the expression $x^2 + px + q$ by completing the square. He correctly does this and gets $(x - 7)^2 - 52$. What are the values of p and q?

2.5 Algebraic fractions

THIS SECTION WILL SHOW YOU HOW TO ...
✓ simplify algebraic fractions
✓ add, subtract, multiply and divide algebraic fractions

KEY WORDS
✓ algebraic fractions   ✓ reciprocal   ✓ cancel
✓ single fraction   ✓ common factor

These five rules are used to simplify **algebraic fractions**.

Cancelling: $\frac{ac}{ad} = \frac{c}{d}$ a is a **common factor** of the numerator and the denominator and so can be removed from both.

Addition: $\frac{a}{b} + \frac{c}{d} = \frac{ad + bc}{bd}$

Subtraction: $\frac{a}{b} - \frac{c}{d} = \frac{ad - bc}{bd}$

Multiplication: $\frac{a}{b} \times \frac{c}{d} = \frac{ac}{bd}$

Division: $\frac{a}{b} \div \frac{c}{d} = \frac{a}{b} \times \frac{d}{c} = \frac{ad}{bc}$ To divide by a fraction, multiply by the **reciprocal**.

EXAMPLE 22

Simplify the expression $\frac{1}{x} + \frac{x}{2y}$.

Using the addition rule: $\frac{1}{x} + \frac{x}{2y} = \frac{(1)(2y) + (x)(x)}{(x)(2y)}$

$$= \frac{2y + x^2}{2xy}$$

EXAMPLE 23

Write $\frac{3}{x-1} - \frac{2}{x+1}$ as a **single fraction** as simply as possible.

$$\frac{3}{x-1} - \frac{2}{x+1} = \frac{3(x+1) - 2(x-1)}{(x-1)(x+1)}$$

Using the subtraction rule:
$$= \frac{3x + 3 - 2x + 2}{(x-1)(x+1)}$$

$$= \frac{x+5}{(x-1)(x+1)}$$

EXAMPLE 24

Simplify fully $\frac{2x^2 + x - 3}{4x^2 - 9}$.

Factorise the numerator and denominator: $\frac{2x^2 + x - 3}{4x^2 - 9} = \frac{(2x+3)(x-1)}{(2x+3)(2x-3)}$

The denominator is the difference of two squares.

Cancel any common factors:
$$= \frac{\cancel{(2x+3)}(x-1)}{\cancel{(2x+3)}(2x-3)}$$

The remaining fraction is the answer:
$$= \frac{x-1}{2x-3}$$

HINTS AND TIPS

You can only cancel whole brackets and not terms within brackets.

EXERCISE 21

TARGETING PASS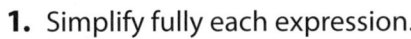

1. Simplify fully each expression.

   a) $\frac{xy}{4} + \frac{2}{x}$

   b) $\frac{2x+1}{2} + \frac{3x+1}{4}$

   c) $\frac{x}{5} + \frac{2x+1}{3}$

   d) $\frac{x-4}{4} + \frac{2x-3}{2}$

   e) $\frac{xy}{4} - \frac{2}{y}$

   f) $\frac{2x+1}{2} - \frac{3x+1}{4}$

   g) $\frac{x}{5} - \frac{2x+1}{3}$

   h) $\frac{x-4}{4} - \frac{2x-3}{2}$

   i) $\frac{4y^2}{9x} \times \frac{3x^2}{2y}$

   j) $\frac{x}{2} \times \frac{x-2}{5}$

   k) $\frac{x-3}{15} \times \frac{5}{2x-6}$

   l) $\frac{x-5}{10} \times \frac{5}{x^2 - 5x}$

   m) $\frac{3x}{4} + \frac{x}{4}$

   n) $\frac{3x}{4} \div \frac{x}{4}$

   o) $\frac{3x+1}{2} - \frac{x-2}{5}$

   p) $\frac{x^2 - 9}{10} \times \frac{5}{x-3}$

2. Write each expression as a single fraction as simply as possible.

a) $\dfrac{2}{x+1} + \dfrac{5}{x+2}$ b) $\dfrac{4}{x-2} + \dfrac{7}{x+1}$ c) $\dfrac{3}{2x-1} - \dfrac{4}{3x-1}$

3. Write each expression as a single fraction.

a) $\dfrac{4}{x+1} + \dfrac{5}{x+2}$ b) $\dfrac{18}{4x-1} - \dfrac{1}{x+1}$ c) $\dfrac{2x-1}{2} - \dfrac{6}{x+1}$

TARGETING MERIT

4. An expression of the form $\dfrac{ax^2 + bx - c}{dx^2 - e}$ simplifies to $\dfrac{x-1}{2x-3}$.
What was the original expression?

5. Simplify fully each expression.

a) $\dfrac{x^2 + 2x - 3}{2x^2 + 7x + 3}$ b) $\dfrac{4x^2 - 1}{2x^2 + 5x - 3}$ c) $\dfrac{6x^2 + x - 2}{9x^2 - 4}$ d) $\dfrac{4x^2 + x - 3}{4x^2 - 7x + 3}$

e) $\dfrac{4x^2 - 25}{8x^2 - 22x + 5}$

6. Simplify fully each expression.

a) $\dfrac{x^2 + 6x - 7}{x^2 - 4} \div \dfrac{x+7}{x^2 + 2x}$ b) $\dfrac{x^2 + x - 30}{x^2 - 16} \div \dfrac{x+6}{x^2 + 4x}$ c) $\dfrac{x^2 - 7x + 12}{x^2 - 25} \div \dfrac{x-3}{x^2 + 5x}$

d) $\dfrac{x^2 + 4x - 60}{x^2 - 49} \div \dfrac{x+10}{x^2 + 7x}$ e) $\dfrac{x^2 - 4x - 5}{x^2 - 9} \div \dfrac{x+1}{x^2 - 3x}$ f) $\dfrac{x^2 - 11x - 42}{x^2 - 64} \div \dfrac{x-14}{x^2 - 8x}$

7. Simplify fully each expression.

a) $\dfrac{x^3 + 3x^2 + 2x}{x^2 + x}$ b) $\dfrac{x^3 + 2x^2 - 3x}{x^2 - x}$ c) $\dfrac{4x^3 + x^2 - 5x}{x^2 - x}$

d) $\dfrac{2x^3 + 3x^2 - 35x}{x^2 + 5x}$ e) $\dfrac{3x^3 + 6x^2 + 3x}{x^2 + x}$ f) $\dfrac{x^3 - 4x}{x^2 - 2x}$

TARGETING DISTINCTION

8. Simplify fully each expression.

a) $\dfrac{6x^2 - 28x - 10}{4x^2 - 9} \div \dfrac{x-5}{4x^2 + 6x}$ b) $\dfrac{9x^2 + 28x - 32}{9x^2 - 49} \div \dfrac{x+4}{3x^2 + 7x}$ c) $\dfrac{3x^2 + 13x + 4}{16x^2 - 9} \div \dfrac{3x+1}{8x^2 + 6x}$

d) $\dfrac{9x^2 - 3x - 6}{25x^2 - 36} \div \dfrac{3x+2}{10x-12}$ e) $\dfrac{10x^2 - 4x - 14}{64x^2 - 9} \div \dfrac{5x-7}{24x^2 + 9x}$ f) $\dfrac{8x^2 - 6x - 5}{9x^2 - 16} \div \dfrac{2x+1}{3x^2 - 4x}$

TARGETING DISTINCTION*

9. $\dfrac{14x^3 - 76x^2 + 30x}{7x^2 + 25x - 12} = \dfrac{ax(x+b)}{x+c}$, where a, b and c are constants.

Work out the values of a, b and c.

Exam-style questions

1. Simplify as much as possible $\dfrac{(6y)^2 + (4y)^2}{6y - 4y}$. [3 marks]

2. Simplify $\dfrac{x^{\frac{1}{2}} \times x^{1\frac{1}{2}}}{x^2 \times x^{-1}}$ as fully as possible. [3 marks]

3. Write $\dfrac{\sqrt{x}}{\sqrt[3]{x}}$ as a power of x. [2 marks]

4. Write 9×27^a as a power of 3 in its simplest form [3 marks]

5. Simplify $\dfrac{3x - x^{\frac{3}{2}} + \sqrt{x}}{\sqrt{x}}$ as fully as possible. [3 marks]

6. Simplify $(2d + 3)(d - 2) - 2(d - 2)^2$. [4 marks]

7. Expand and simplify.
   a) $(4 - x)^3$ [4 marks]
   b) $x^{\frac{1}{2}}\left(x^{\frac{1}{2}} + x^{-\frac{1}{2}}\right)$ [2 marks]

8. The coefficient of x^3 in the expansion of $(2x - a)^5$ is 720, where a is a positive integer.
   Use Pascal's triangle to show that $a = 3$. [3 marks]

9. Prove that $(4x + 1)^2 - 7x(x + 1) + 5x$ is always greater than or equal to 0. [4 marks]

10. a) Factorise $9x^2 - 4$. [1 mark]
    b) Factorise $6x^2 - x - 2$. [2 marks]

11. Factorise completely $(5x - 3)^2 - (3x - 5)^2$. [3 marks]

12. Factorise completely $-12x^4 - 30x^3 + 72x^2$. [3 marks]

13. Find the values of a, b and c such that $6x^2 + 6x - 1 = a(x + b)^2 + c$. [3 marks]

14. Write $\dfrac{2x + 1}{x + 2} - \dfrac{2x - 1}{x + 1}$ as a single fraction, as simply as possible. [4 marks]

15. Simplify fully $\dfrac{4x^3 - x}{x - 3} \div \dfrac{4x^3 + 4x^2 + x}{x^2 - 6x + 9}$. [4 marks]

16. Simplify fully $\dfrac{x^3 - 4x^2 + 4x}{x^4 - 8x^2 + 16}$. [4 marks]

17. $f(x) = x^3 - x^2 - 14x + 24$
    a) Show that $x + 4$ is a factor of $f(x)$. [2 marks]
    b) Factorise $f(x)$. [3 marks]

18. Fully factorise $x^3 - 4x^2 - 25x + 100$. [4 marks]

19. a) Show that $2x + 1$ is a factor of $2x^3 - x^2 - 13x - 6$. [2 marks]
    b) Hence, fully factorise $2x^3 - x^2 - 13x - 6$. [4 marks]

3 Graphs

3.1 Linear graphs

THIS SECTION WILL SHOW YOU HOW TO ...
- ✓ draw straight-line graphs
- ✓ find the equation of a straight line
- ✓ find equations of parallel and perpendicular lines
- ✓ find equations of tangents of circles

KEY WORDS
- ✓ constant term
- ✓ negative reciprocal
- ✓ tangent
- ✓ gradient
- ✓ parallel
- ✓ intercept
- ✓ perpendicular

Equations of linear graphs can be written in the following forms:
- $y = mx + c$, e.g. $y = 3x + 5$
- $ax + by = c$ or $ax + by + c = 0$, e.g. $3x - y = -5 \Rightarrow 3x - y + 5 = 0$
- $y - y_1 = m(x - x_1)$, e.g. $y - 8 = 3(x - 1)$ using the coordinate $(1, 8)$ and a gradient of 3.

The different forms have different uses and properties.

When a graph is expressed in the form $y = mx + c$, the coefficient of x, m, is the **gradient**, and the **constant term**, c, gives the **intercept** on the y-axis.

The gradient of a line can be calculated by dividing the 'difference in the y-coordinates', Δy, by the 'difference in the x-coordinates', Δx.

$$m = \frac{\text{difference in the } y\text{-coordinates}}{\text{difference in the } x\text{-coordinates}} = \frac{\Delta y}{\Delta x}$$

The y-intercept of a line is where the graph meets the y-axis.

EXAMPLE 1

Rearrange $y = -\frac{2}{3}x - 4$ into the form $ax + by + c = 0$.

$$y = -\frac{2}{3}x - 4$$

Multiply the equation by 3: $3y = -2x - 12$
Rearrange: $\quad 2x + 3y + 12 = 0$

Finding the equation of a straight line

If you know the gradient, m, of a line and its intercept, c, on the y-axis, you can write down the equation of the line immediately.

For example, if $m = 3$ and $c = -5$, the equation of the line is $y = 3x - 5$.

All linear graphs can be expressed in the form $y = mx + c$.

There are several methods for finding the equation of a straight line based on the information you are given.

EXAMPLE 2

Find the equation of the straight line shown in diagram **A**.

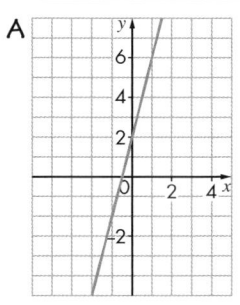 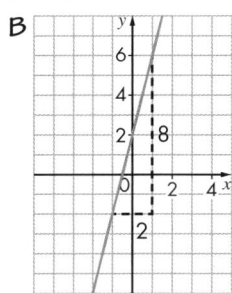

First, find where the graph crosses the y-axis.
This is at $(0, 2)$ so $c = 2$.
Next, calculate the gradient (diagram **B**)
$m = \frac{\Delta y}{\Delta x} = \frac{8}{2} = 4$

Finally, write down the equation of the line: $y = 4x + 2$

Instead of a graph, you may be given different information about a straight line. For example:

- the gradient, m, and the coordinates of one point on the line (x_1, y_1)
- the coordinates of two points on the line (x_1, y_1) and (x_2, y_2).

If you are given the gradient and the coordinates of one point on the line, you can substitute these into $y = mx + c$ to find the y-intercept, or use the form $y - y_1 = m(x - x_1)$.

If you are given the coordinates of two points on the line, to find the gradient you can work out the difference between the y-coordinates and between the x-coordinates and divide:

$m = \frac{\text{difference in the } y\text{-coordinates}}{\text{difference in the } x\text{-coordinates}} = \frac{\Delta y}{\Delta x} = \frac{y_2 - y_1}{x_2 - x_1}$

EXAMPLE 3

Find the equation of the straight line passing through the points $(-1, -2)$ and $(1, 6)$.

$m = \frac{\text{difference in the } y\text{-coordinates}}{\text{difference in the } x\text{-coordinates}} = \frac{\Delta y}{\Delta x} = \frac{y_2 - y_1}{x_2 - x_1} = \frac{6 - -2}{1 - -1} = \frac{8}{2} = 4$

Using $y = mx + c$

$y = 4x + c$
$(6) = 4(1) + c$
$6 = 4 + c$
$c = 2$

Using $y - y_1 = m(x - x_1)$

$y - y_1 = 4(x - x_1)$
$y - 6 = 4(x - 1)$
$y - 6 = 4x - 4$
$y = 4x + 2$

Therefore, $y = 4x + 2$

Alternatively, you can use the other pair of coordinates – it doesn't *matter* which you use.

Parallel and perpendicular lines

If two lines are **parallel**, then their gradients are equal.

If two lines are **perpendicular** (at right angles), then their gradients are **negative reciprocals** of each other.

> **HINTS AND TIPS**
>
> If one line has gradient m_1 and the other has gradient m_2, then, if the lines are perpendicular, $m_1 \times m_2 = -1$.
> Therefore, if $m_1 = 2$
> $2 \times m_2 = -1 \Rightarrow m_2 = -\frac{1}{2}$

EXAMPLE 4

For each grid:
   i) find the equation of each line
  ii) describe the numerical relationships between the gradients of the lines
 iii) describe the geometrical relationship between the lines.

a)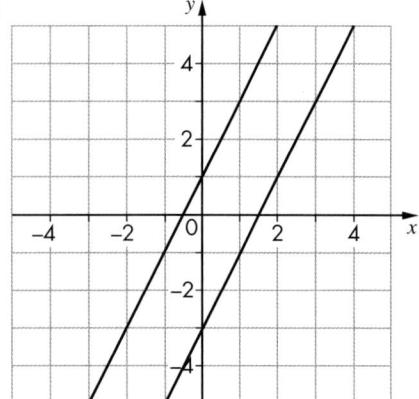

b)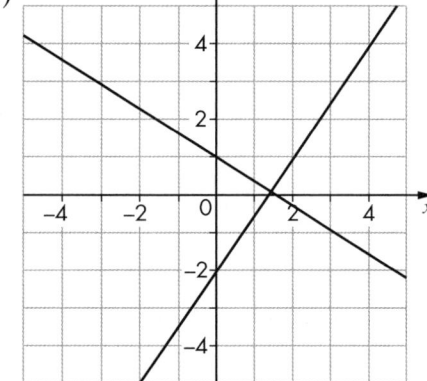

Grid **a**:
  i) The lines have equations
     $y = 2x + 1$, $y = 2x - 3$.
  ii) The gradients are equal.
 iii) The lines are parallel.

Grid **b**:
  i) The lines have equations
     $y = \frac{3}{2}x - 2$, $y = -\frac{2}{3}x + 1$.
  ii) The gradients are reciprocals of each other but with different signs:
     $\frac{3}{2} \times -\frac{2}{3} = -1$.
 iii) The lines are perpendicular.

EXAMPLE 5

Two points on a straight line have coordinates A(0, 1) and B(2, 4).
a) Work out the equation of the line AB.
b) Write down the equation of the line parallel to AB that passes through the point (0, 5).
c) Write down the gradient of a line perpendicular to AB.
d) Write down the equation of a line perpendicular to AB that passes through the point (0, 2). Give your answer in the form $ax + by + c = 0$.

a) The gradient of AB is $\frac{4-1}{2-0} = \frac{3}{2}$

The line passes through (0, 1), which is the y-intercept, so the equation is $y = \frac{3}{2}x + 1$.

b) Parallel lines have equal gradients, so a line parallel to AB will have gradient $\frac{3}{2}$.

The point (0, 5) is the y-intercept, so the equation is $y = \frac{3}{2}x + 5$.

c) The gradient of a line perpendicular to AB is the negative reciprocal of $\frac{3}{2}$, which is $-\frac{2}{3}$.

d) The gradient is $-\frac{2}{3}$ and the point (0, 2) is the y-intercept, so the equation is
$y = -\frac{2}{3}x + 2$.

Multiply both sides by 3: $y = -\frac{2}{3}x + 2$

Rearrange: $\quad\quad\quad\quad 3y = -2x + 6$
$\quad\quad\quad\quad\quad\quad 2x + 3y - 6 = 0$

This is now in the form $ax + by + c = 0$.

EXAMPLE 6

Line A has the equation $y = \frac{1}{2}x - 3$.
Line B is perpendicular to line A and passes through (1, 5). Find the equation of line B.

The gradient of line B is the negative reciprocal of $\frac{1}{2}$ which is −2.

Using $y = mx + c$
$y = -2x + c$
$5 = -2(1) + c$
$5 = -2 + c$
$c = 7$
Therefore, $y = -2x + 7$

Using $y - y_1 = m(x - x_1)$
$y - y_1 = -2(x - x_1)$
$y - 5 = -2(x - 1)$
$y - 5 = -2x + 2$
$y = -2x + 7$

EXAMPLE 7

The point A is (2, 1) and the point B is (4, 4).
Find the equation of the line parallel to AB that passes through (6, 11).

First, calculate the gradient of AB.

$m = \frac{\text{difference in the y-coordinates}}{\text{difference in the x-coordinates}} = \frac{\Delta y}{\Delta x} = \frac{y_2 - y_1}{x_2 - x_1} = \frac{4-1}{4-2} = \frac{3}{2}$

The line parallel to AB has the same gradient, $\frac{3}{2}$.

Using $y = mx + c$

$y = \frac{3}{2}x + c$

$11 = \frac{3}{2}(6) + c$

$11 = 9 + c$

$c = 2$

Using $y - y_1 = m(x - x_1)$

$y - y_1 = \frac{3}{2}(x - x_1)$

$y - 11 = \frac{3}{2}(x - 6)$

$y - 11 = \frac{3}{2}x - 9$

$y = \frac{3}{2}x + 2$

Therefore, $y = \frac{3}{2}x + 2$

EXAMPLE 8

The point A has coordinates (2, −1) and the point B has coordinates (4, 5).

Find the equation of the perpendicular bisector of AB.

First, calculate the gradient of AB.

$m = \dfrac{\text{difference in the } y\text{-coordinates}}{\text{difference in the } x\text{-coordinates}} = \dfrac{\Delta y}{\Delta x} = \dfrac{y_2 - y_1}{x_2 - x_1}$

$= \dfrac{5 - -1}{4 - 2} = \dfrac{6}{2} = 3$

HINTS AND TIPS

You can find the coordinates of the midpoint of a line joining points (x_1, y_1) and (x_2, y_2) by using $\left(\dfrac{x_1 + x_2}{2}, \dfrac{y_1 + y_2}{2}\right)$.

The gradient of a line perpendicular to AB is the negative reciprocal of 3, which is $-\frac{1}{3}$.

To bisect AB, the perpendicular line must pass through the midpoint, which has coordinates

$\left(\dfrac{2 + 4}{2}, \dfrac{-1 + 5}{2}\right) = \left(\dfrac{6}{2}, \dfrac{4}{2}\right) = (3, 2)$

Then use the general equation of a straight line: $y - y_1 = m(x - x_1)$

Substitute the gradient and coordinates of the midpoint: $y - 2 = -\frac{1}{3}(x - 3)$

Rearrange: $y - 2 = -\frac{1}{3}x + 1$

$y = -\frac{1}{3}x + 3$

Circles and tangents

A **tangent** to a curve is a point that touches it at one point only.

A tangent to a circle is perpendicular to the radius at the point of contact.

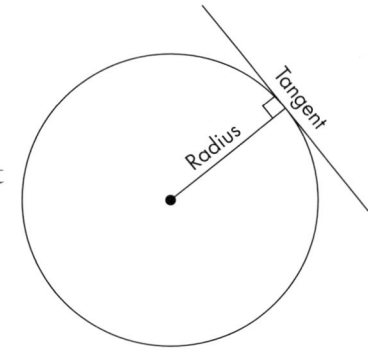

EXAMPLE 9

A circle with centre $(2, 3)$ has a tangent at the point $(6, 4)$. Find the equation of the tangent.

HINTS AND TIPS

A sketch helps to visualise the problem.

Find the gradient of the radius

$$m = \frac{\Delta y}{\Delta x} = \frac{y_2 - y_1}{x_2 - x_1} = \frac{4 - 3}{6 - 2} = \frac{1}{4}$$

Find the gradient of the tangent

As the radius and the tangent are perpendicular, the gradient of the tangent is the negative reciprocal of the gradient of the radius.

$m_2 = -4$

Find the equation of the tangent

$y - 4 = -4(x - 6)$ (using $y - y_1 = m(x - x_1)$)

$y - 4 = -4x + 24$

$y = -4x + 28$

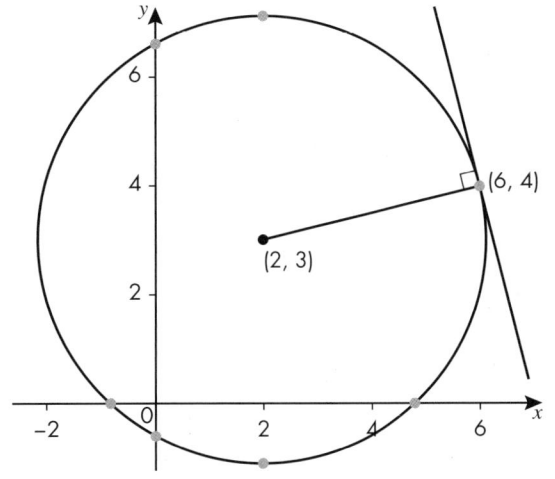

EXERCISE 3A

TARGETING PASS

1. Find the equation of the line joining each pairs of points. Give your answers in the form $y = mx + c$.

    a) $(0, -3)$ and $(4, 5)$ **b)** $(-4, 2)$ and $(2, 5)$

2. Rearrange your answers to question **1** into the form $ax + by + c = 0$.

3. Find an expression for the gradient of the line joining the points $(3p, -7q)$ and $(p - 2, 4 - q)$. Give your answer in terms of p and q.

4. Write down the equations of the line that is:

    a) parallel to $y = 4x - 5$ that passes through $(0, 1)$

    b) parallel to $y = \frac{1}{2}x + 3$ that passes through $(0, -2)$.

5. Write down the equations of the line that is:

    a) perpendicular to $y = 3x + 2$ that passes through $(0, -1)$

    b) perpendicular to $y = \frac{1}{3}x - 2$ that passes through $(0, 5)$.

6. For each of these lines, write down the equation of the perpendicular line that passes through the same point on the y-axis.

    a) $y = 2x - 1$ **b)** $y = -3x + 1$ **c)** $y = \frac{1}{2}x + 3$ **d)** $y = -\frac{2}{3}x - 5$

TARGETING MERIT

7. Find the equation of the line perpendicular to $y = 4x - 3$ that passing though the point $(-4, 3)$.

8. A is the point $(1, 5)$ and B is the point $(3, 3)$.
   a) Find the equation of the line parallel to AB that passes through $(5, 9)$.
   b) Find the equation of the perpendicular bisector of AB.

   HINTS AND TIPS
   Remember that the bisector of a line passes through its midpoints.

9. Find the equation of the perpendicular bisector of the line joining the points A$(1, 2)$ and B$(3, 6)$.

10. Line A passes through the points $(-3, 0)$ and $(4, 3)$.
    Line B passes through the points $(1, -6)$ and $(-13, -12)$.
    Show that line A and line B are parallel.

11. Two lines, l_1 and l_2, are perpendicular.
    The equation of l_1 is $3x + y = 10$.
    l_1 and l_2 cross at the point $(3, 5)$.
    Work out the equation of l_2.

12. A circle with centre $(0, 0)$ has a tangent, l, at the point $(-2, 3)$. Find the equation of l.

TARGETING DISTINCTION

13. Line C passes through the points $(2, -5)$ and $(5, 7)$.
    Line D passes through the points $(1, 3)$ and $(-7, p)$.
    Given that line C and line D are perpendicular, find the value of p.

14. A is the point $(3, 4)$, B is the point $(7, 2)$, C is the point $(10, -3)$ and D is the point $(2, 1)$. What type of quadrilateral is ABCD? Explain how you decide.

    HINTS AND TIPS
    Find the gradient of each side.

15. A quadrilateral has its vertices at A$(4, 6)$, B$(6, -2)$, C$(-6, -5)$ and D$(-8, 3)$.
    Show that ABCD is a rectangle.

    HINTS AND TIPS
    A rectangle has two pairs of parallel sides and four right angles. In **Q15**, use gradients to prove that ABCD has these.

16. A circle with centre $(-2, 5)$ has a tangent, l, at the point $(3, -7)$. Find the equation of l, in the form $ax + by = c$.

17. A circle with centre $(0, 0)$ meets the line $2x + y = 3$ at the point $\left(\frac{2}{5}, \frac{11}{5}\right)$. Determine whether $2x + y = 3$ is a tangent to the circle.

TARGETING DISTINCTION*

18. A is the point $(-2, p)$, B is the point $(-6, 7)$ and C is the point $(q, 3)$.

Given that angle ABC is a right angle, show that $p - q = 13$.

> **HINTS AND TIPS**
>
> Find the gradients of lines AB and BC.

19. The line l_1 passes through the points $P(-2, 1)$ and $Q(12, 7)$

   a) Find the equation for l_1 in the form $ax + by = c$, where a, b and c are constants.

   The line l_2 passes through the point $R(9, n)$ and is perpendicular to l_1.

   b) Write an expression for n in terms of x and y.

   c) Given that l_2 also passes through point P, find the equation of l_2 in the form $ax + by = c$.

3.2 Quadratic graphs

THIS SECTION WILL SHOW YOU HOW TO ...
✓ recognise and calculate the significant points of a quadratic graph
✓ sketch quadratic graphs
✓ understand the discriminant

KEY WORDS
✓ discriminant
✓ root(s)
✓ maximum/minimum
✓ turning point
✓ parabola
✓ vertex

Quadratic graphs are diagrams that represent quadratic functions.

A quadratic function contains an x^2 term as the highest power of x: for example, $y = x^2 + 3x + 2$, $y = x^2 + 3x$ or $y = 2x^2 + 3x - 2$

The general form of a quadratic graph is $y = ax^2 + bx + c$.

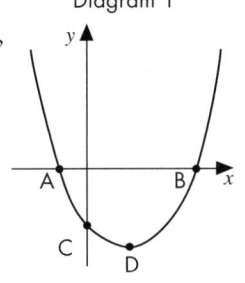

Diagram 1

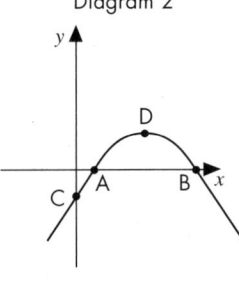

Diagram 2

A quadratic graph has four points that define its shape. These are the points A, B, C and D on the diagram. The x-values at A and B where the graph crosses the x-axis (the x-intercepts) are the **roots** of the equation $ax^2 + bx + c = 0$. C is the point where the graph crosses the y-axis (the y-intercept) and D is the **vertex** or **turning point**, which can be a **maximum** (as in diagram 1) or a **minimum** (as in diagram 2) point. A quadratic graph has a vertical line of symmetry, which passes through the vertex.

Quadratic curves are called **parabolas**. Diagram 1 shows a positive quadratic graph for which the x^2 term is positive ($a > 0$). Diagram 2 shows a negative quadratic graph for which the x^2 term is negative ($a < 0$).

The roots
There are four methods that you can use to find the roots of a quadratic equation.

EXAMPLE 10

Determine the roots of the equation $x^2 - 2x - 3 = 0$.

Using a quadratic graph

Draw the graph of $y = x^2 - 2x - 3$. The roots of the equation can be found where the parabola crosses the x-axis (the x-intercepts).

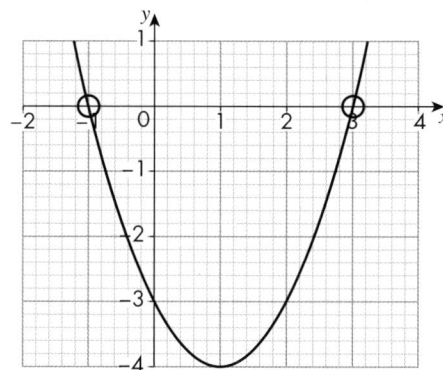

HINTS AND TIPS
The x-axis has equation $y = 0$.

By factorising
$x^2 - 2x - 3 = 0$
$(x - 3)(x + 1) = 0$
$x = -1, x = 3$

Therefore, the roots are $x = -1$, $x = 3$.

HINTS AND TIPS
Factorising is the method that you should usually try first when solving quadratic equations.

Using the quadratic formula
$x^2 - 2x - 3 = 0$
$a = 1, b = -2, c = -3$
$x = \dfrac{-b \pm \sqrt{b^2 - 4ac}}{2a}$
$x = \dfrac{-(-2) \pm \sqrt{(-2)^2 - 4(1)(-3)}}{2(1)}$
$= \dfrac{2 \pm \sqrt{4 + 12}}{2}$
$= \dfrac{2 \pm 4}{2}$
$x = -1, x = 3$

HINTS AND TIPS
You need to learn this formula.

By completing the square
$x^2 - 2x - 3 = 0$
$(x - 1)^2 - 1 - 3 = 0$
$(x - 1)^2 - 4 = 0$
$(x - 1)^2 = 4$
$x - 1 = \pm\sqrt{4} = \pm 2$
$x = -1, x = 3$

HINTS AND TIPS
You learned how to complete the square in Section 2.4.

The y-intercept

If you look at all the quadratic graphs presented so far you will see a connection between the equation and the point where the graph crosses the y-axis. The constant term of the equation $y = ax^2 + bx + c$ (that is, the value c) is where the graph crosses the y-axis. The intercept is at $(0, c)$.

The vertex

The lowest (or highest) point of a quadratic graph is called the vertex or turning point.

If the x^2 term is negative ($a < 0$), then the vertex is a maximum.

If the x^2 term is positive ($a > 0$), then the vertex is a minimum.

There are three methods for finding the vertex of a quadratic graph.

EXAMPLE 11

Determine the coordinates of the vertex of the graph of $y = x^2 - 2x - 3$.

As the x^2 term is positive ($a > 0$), $y = x^2 - 2x - 3$ will have a minimum point.

Using the quadratic graph

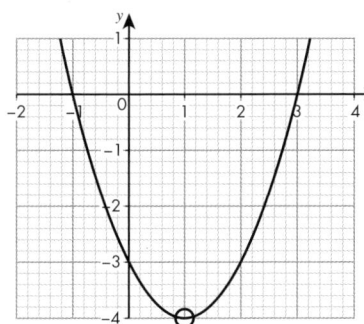

The vertex of the graph is at $(1, -4)$.

HINTS AND TIPS

When the expression is written in completed square form, the x-coordinate of the vertex is the value of x that makes the bracket equal to zero and the y-coordinate is the final constant.

By completing the square

$x^2 - 2x - 3$
$(x - 1)^2 - 1 - 3$
$(x - 1)^2 - 4$

The x-coordinate of the vertex is the value of x that makes $x - 1 = 0$. Therefore, $x = 1$.

This is because the lowest value that $(x \pm a)^2$ can take is zero and to find the minimum point, you must minimise the value of $x^2 - 2x - 3$.

If $x = 1$,
$(x - 1)^2 - 4 = ((1) - 1)^2 - 4$
$= -4$

Therefore, the y-coordinate of the vertex is -4.

The vertex of the graph is at $(1, -4)$.

Using x-intercepts and substitution

Because of the symmetry of the curve, the x-coordinate of the vertex will always be halfway between the x-intercepts of the quadratic graph.

In **Example 12**, you determined the roots of $x^2 - 2x - 3 = 0$ to be $x = -1$, $x = 3$. So these are the x-intercepts.

Halfway between -1 and 3 is 1.

To calculate the y-coordinate, you can substitute $x = 1$ into $x^2 - 2x - 3$.

$(1)^2 - 2(1) - 3 = 1 - 2 - 3 = -4$

The vertex of the graph is at $(1, -4)$.

HINTS AND TIPS

Although this seems more straightforward than completing the square, you need to know the x-intercepts of the graph, which may not always be necessary or possible.

EXAMPLE 12

Sketch the graph of $y = x^2 + 3x + 2$.

As $a > 0$, the curve is $\smile$ shaped.

Roots
$x^2 + 3x + 2 = 0$
$(x + 2)(x + 1) = 0$
Roots are $x = -2, -1$, so the x-intercepts are $(-2, 0)$ and $(-1, 0)$.

y-intercept
$y = x^2 + 3x \boxed{+ 2}$
y-intercept is $(0, 2)$

Vertex
$x^2 + 3x + 2$
$= \left(x + \frac{3}{2}\right)^2 - \frac{9}{4} + 2$
$= \left(x + \frac{3}{2}\right)^2 - \frac{1}{4}$
Vertex is $\left(-\frac{3}{2}, -\frac{1}{4}\right)$

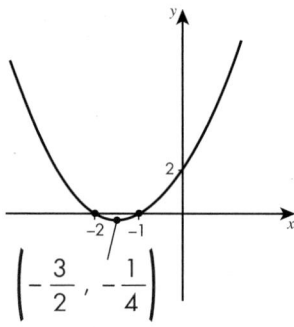

EXAMPLE 13

Sketch the graph of $y = 2x^2 + 7x + 5$.

As $a > 0$, the curve is $\smile$ shaped.

Roots
$2x^2 + 7x + 5 = 0$
$(2x + 5)(x + 1) = 0$
Roots are $x = -\frac{5}{2}, -1$, so the x-intercepts are $\left(-\frac{5}{2}, 0\right)$ and $(-1, 0)$.

y-intercept
$y = 2x^2 + 7x \boxed{+ 5}$
y-intercept is $(0, 5)$

Vertex
$2x^2 + 7x + 5$
$= 2\left[x^2 + \frac{7}{2}x\right] + 5$
$= 2\left[\left(x + \frac{7}{4}\right)^2 - \frac{49}{16}\right] + 5$
$= 2\left(x + \frac{7}{4}\right)^2 - \frac{49}{8} + 5$
$= 2\left(x + \frac{7}{4}\right)^2 - \frac{9}{8}$
Vertex is $\left(-\frac{7}{4}, -\frac{9}{8}\right)$

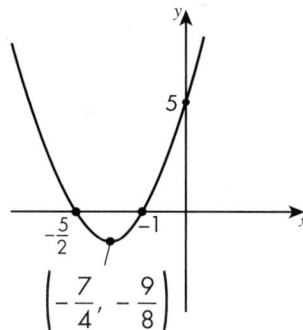

EXERCISE 3B

TARGETING PASS

1. Determine the *x*-intercepts, *y*-intercept and vertex of each of these quadratic graphs.

a) 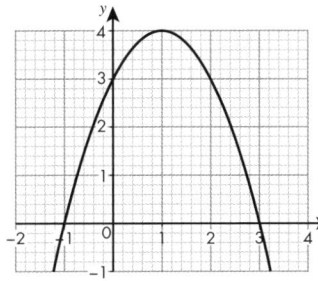   b) 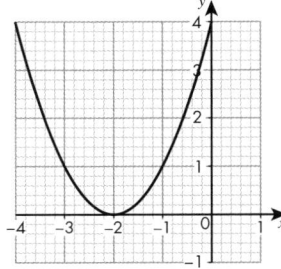   c)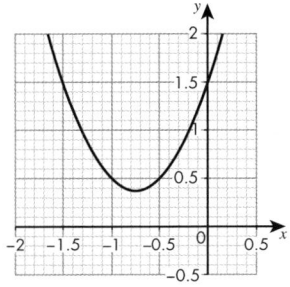

HINTS AND TIPS

In **c)**, where the graph does not cross the *x*-axis. You say that the equation $y = 0$ has 'no real roots'.

TARGETING DISTINCTION

2. Sketch the graphs of these quadratic functions, by finding:
  i) the *x*-intercepts by solving the equation $y = 0$ by factorising
  ii) the *y*-intercept
  iii) the vertex by completing the square.

  a) $y = x^2 + 6x + 8$ **b)** $y = x^2 + 10x + 24$ **c)** $y = x^2 - 7x + 12$ **d)** $y = x^2 + 3x - 4$

TARGETING DISTINCTION*

3. Sketch the graphs of these quadratic functions.
  a) $y = 4x^2 + 8x + 3$
  b) $y = 2x^2 + 5x + 2$
  c) $y = -x^2 + 10x - 25$
  d) $y = -2x^2 - 17x - 36$

4. Using the quadratic graphs, determine the values of a, b and c and give the equation in the form $y = ax^2 + bx + c$.

a) 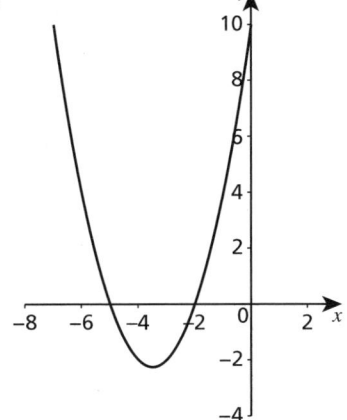   b) 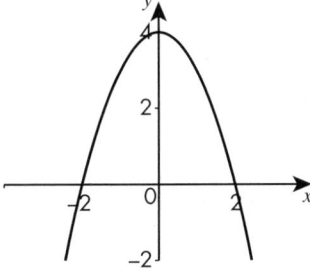   c)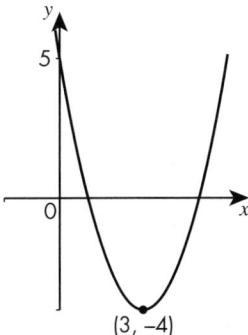

The discriminant

You can determine how many roots a quadratic equation has by considering the **discriminant**.

$x = \dfrac{-b \pm \sqrt{b^2 - 4ac}}{2a}$ In the quadratic formula, $b^2 - 4ac$ is the discriminant.

- If $b^2 - 4ac > 0$, the quadratic graph has two real roots.
- If $b^2 - 4ac = 0$, the quadratic graph has one real (two equal) roots.
- If $b^2 - 4ac < 0$, the quadratic graph has no real roots.

This is because the discriminant is the expression under the square root in the quadratic formula. As you cannot find the square root of a negative number, if $b^2 - 4ac < 0$, there are no real solutions to the equation $ax^2 + bx + c = 0$.

Two real roots: $b^2 - 4ac > 0$

The graph crosses the x-axis at two points.

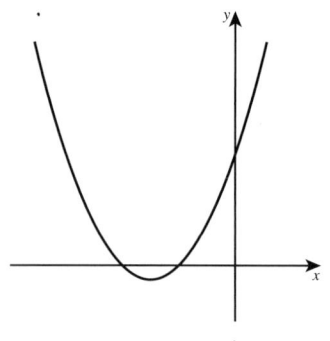
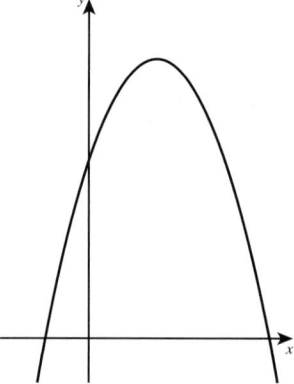

One real root: $b^2 - 4ac = 0$; also called equal roots.

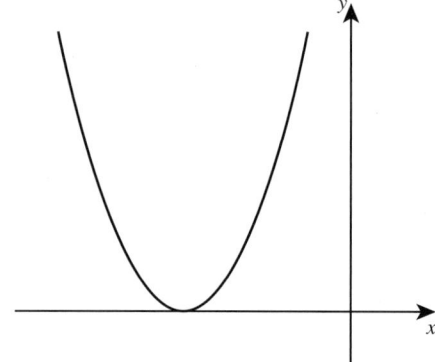
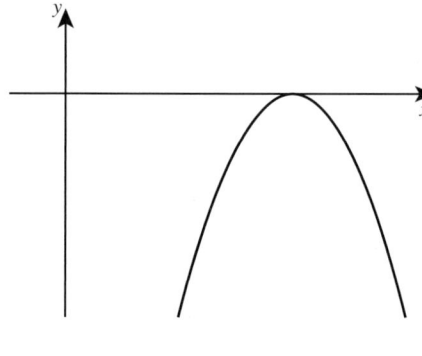

No real roots: $b^2 - 4ac < 0$

The graph does not meet the x-axis.

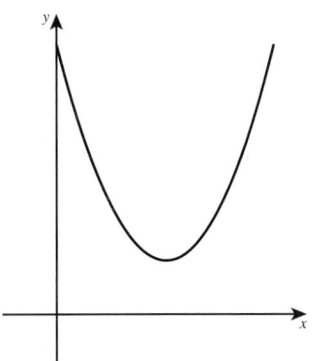
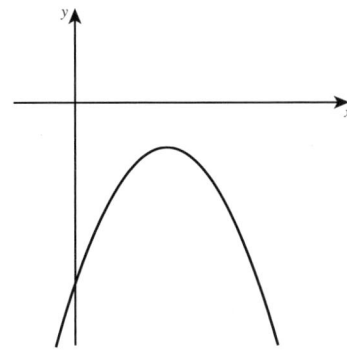

EXAMPLE 14

Determine the number of roots of the quadratic equation $2x^2 + 12x + 18 = 0$.

$b^2 - 4ac = (12)^2 - 4(2)(18) = 144 - 144 = 0$
Therefore, the equation has one real root.

EXAMPLE 15

The equation $ax^2 + 5x + 1 = 0$ has two real roots.
Find the largest integer value of a.

$b^2 - 4ac > 0$, as there are two real roots.
$(5)^2 - 4a(1) > 0 \Rightarrow 25 - 4a > 0 \Rightarrow 4a < 25 \Rightarrow a < \frac{25}{4}$
$a < 6\frac{1}{4}$. So the largest integer value of a is 6.

EXAMPLE 16

$y = 2x^2 + bx + 9$ has two equal roots.
Find the positive value of b. Give your answer in the form $k\sqrt{2}$, where k is an integer.

$b^2 - 4ac = 0$, as the equation has two equal roots.
$b^2 - 4(2)(9) = 0 \Rightarrow b^2 - 72 = 0 \Rightarrow b^2 = 72 \Rightarrow b = \pm\sqrt{72} = \pm\sqrt{36}\sqrt{2} = \pm 6\sqrt{2}$
The positive value of b is $6\sqrt{2}$.

EXERCISE 3C

TARGETING DISTINCTION*

1. Determine the number of roots for the following quadratic graphs by calculating the discriminant.

   a) $y = x^2 - 6x + 9$ b) $y = x^2 - 10x - 24$ c) $y = x^2 - 7x + 40$ d) $y = 2x^2 + 6x - 8$

   e) $y = x^2 + 6$ f) $y = 3x^2 + 24x + 48$ g) $y = -x^2 + 9x - 26$ h) $y = -2x^2 - 7x + 12$

2. The following quadratic graphs have two real roots. Find the possible range of values of p.

   a) $y = px^2 + 23x + 6$ b) $y = 2x^2 + 7x + p$

   c) $y = x^2 - 4x + 3p$

   > **HINTS AND TIPS**
   >
   > Remember that 'two real roots' means that the discriminant is greater than zero.

3. The following quadratic graphs have equal roots. Find the possible values of q, where $q > 0$.

   a) $y = x^2 + qx + 4$ b) $y = qx^2 + 12x + 9q$ c) $y = 3x^2 - qx + \frac{1}{2}q$

4. Given that $y = (n-1)x^2 + 3x - n^2$ has no real roots, show that $n^2(n-1) < -\frac{9}{4}$.

3.3 Cubic and quartic graphs

THIS SECTION WILL SHOW YOU HOW TO ...
✓ recognise and calculate the significant points of cubic and quartic graphs
✓ sketch cubic and quartic graphs

KEY WORDS
✓ Local maxima/minima

Cubic graphs

A cubic function contains an x^3 term as the highest power of x: for example, $y = x^3 + x^2 - 3x + 2$, $y = x^3 + 2x$, $y = -2x^3 - 3x^2 + 3x - 2$.

The general form of a cubic graph is $y = ax^3 + bx^2 + cx + d$, where $a > 0$.

You can sketch the graphs of cubic functions by considering where it meets or crosses the x-axis (points A, B and C on the following diagrams) and its y-intercept (where it crosses the y-axis, that is, point D). A cubic graph usually (although not always) has two vertices or turning points (points E and F). These are called **local minima/local maxima**. When asked to sketch a cubic graph, you do not need to know the exact locations of these points.

Diagram 1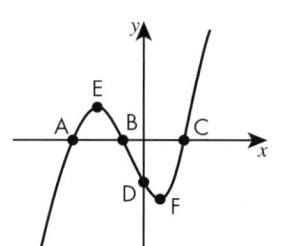

Diagram 2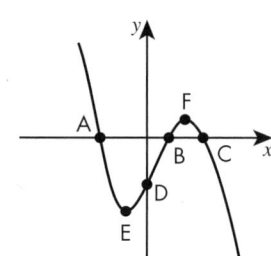

Diagram 1 shows a positive cubic function for which the x^3 term is positive ($a > 0$). Diagram 2 shows a negative cubic function for which the x^3 term is negative ($a < 0$).

To sketch a cubic function, first write it in factorised form. You can then identify where the graph meets or crosses the x-axis by solving the equation $y = 0$.

HINTS AND TIPS

The x-coordinates of the points of intersection with the x-axis are the roots of the equation $ax^3 + bx^2 + cx + d = 0$.

EXAMPLE 17

Sketch the graph of the function $y = (x - 3)(x - 1)(x + 2)$.

Shape
Determine whether the function is positive or negative by finding the coefficient of x^3.
$x \times x \times x = x^3$
$a = 1$. As $a > 0$, the curve has this shape: $\mathcal{N}$

Roots
When $(x - 3)(x - 1)(x + 2) = 0$
$x = 3$, $x = 1$, $x = -2$, so the curve intersects the x-axis at $(3, 0)$, $(1, 0)$ and $(-2, 0)$.

y-intercept
There are two methods for find the y-intercept.

Using $x = 0$
The y-intercept is where the graph intersects the y-axis. This is the line $x = 0$. Therefore, you can substitute $x = 0$ into
$y = (x - 3)(x - 1)(x + 2)$ to find the y-intercept.
$y = ((0) - 3)((0) - 1)((0) + 2)$
$y = (-3) \times (-1) \times 2 = 6$
Therefore, the cubic curve intersects the y-axis at $(0, 6)$.

By expanding brackets
Expanding brackets in the form $(x + p)(x + q)(x + r)$, gives the general cubic form $ax^3 + bx^2 + cx + d$. The constant term, d, is found by multiplying p, q and r.

$pqr = d$
$(-3) \times (-1) \times 2 = 6$
Therefore, the cubic curve intersects the y-axis at $(0, 6)$.

Local maxima/minima
Although you do not need to know the exact coordinates of these turning points, the x-coordinate will be roughly half way between the points of intersection with the x-axis.

Now you can sketch the curve going through the points of intersection with the x-axis and the y-axis.

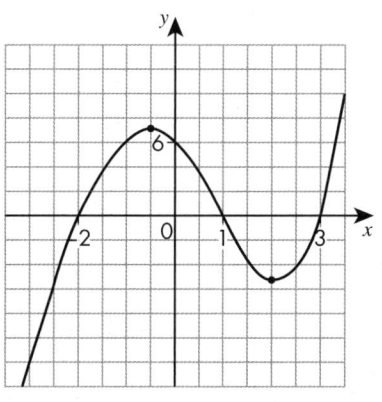

EXAMPLE 18

Sketch the graph of the function $y = x(2x + 5)(3 - x)$.

Shape

$x \times 2x \times (-x) = -2x^3$

$a = -2$. As $a < 0$, the curve has this shape:

Roots

$x(2x + 5)(3 - x) = 0$

The roots are $x = 0$, $x = -\frac{5}{2}$, $x = 3$, so the x-intercepts are $(0, 0)$, $\left(-\frac{5}{2}, 0\right)$ and $(3, 0)$.

y-intercept

The factor x means that there is no constant term, d, for this function. Therefore, the curve intersects the y-axis at the origin $(0, 0)$. You can show this using the two methods given in **Example 17**.

Using $x = 0$

$y = (0)(2(0) + 5)(3 - (0))$
$y = 0 \times 5 \times 3 = 0$

By expanding brackets

$pqr = d$
$0 \times 5 \times 3 = 0$

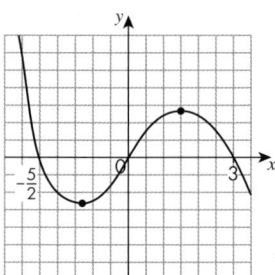

In Examples 17 and 18, the factors of each function $(x - 3)$, $(x - 1)$ and $(x + 2)$ (**Example 17**) and x, $(2x + 5)$ and $(3 - x)$ (**Example 18**) are linear. However, the factors of a cubic function could also be quadratic, for example, $(x + p)^2$ or cubic, for example, $(x + p)^3$. These influence how you sketch the curve.

Linear: $x + p$

The curve crosses the x-axis at the determined points. You have seen these in **Example 17** and **Example 18**.

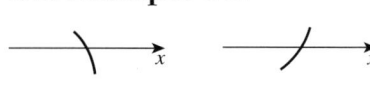

Quadratic: $(x + p)^2$

The curve meets or touches the x-axis at the determined point.

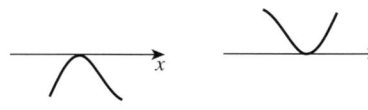

This is called a repeated root.

Cubic: $(x + p)^3$

The curve crosses the x-axis and then continues in the same direction. However, the curve changes from being convex to concave (or vice-versa).

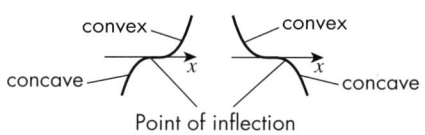

Point of inflection

This is called point of inflection.

EXAMPLE 19

Sketch the graph of the function $y = (x + 2)(x - 3)^2$.

Shape

$x \times x \times x = x^3$

$a = 1$. As $a > 0$, the curve has this shape: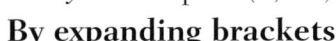

Roots

$(x + 2)(x - 3)^2 = 0$

The roots are $x = -2$, $x = 3$.

There is a repeated root at $x = 3$.

So the graph cuts the x-axis at $(-2, 0)$ and touches it x-axis at $x = 3$.

y-intercept

Using $x = 0$

$y = ((0) + 2)((0) - 3)^2$

$y = 2 \times (-3)^2 = 2 \times 9 = 18$

The y-intercept is $(0, 18)$,

By expanding brackets

$pqr = d$

$2 \times (-3)^2 = 18$

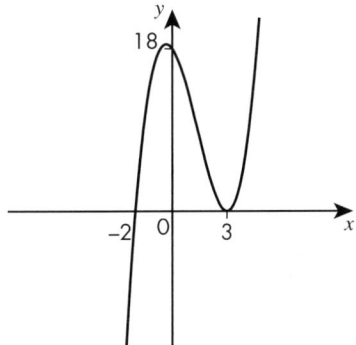

EXAMPLE 20

Sketch the graph of the function $y = (x + 4)^3$.

Shape

$x \times x \times x = x^3$

$a = 1$. As $a > 0$, the curve has this shape: N

Roots

$(x + 4)^3 = 0$

Repeated root at $x = -4$.

There is a point of inflection at $x = -4$.

y-intercept

Using $x = 0$

$y = ((0) + 4)^3$

$y = (4)^3 = 64$

The y-intercept is $(0, 64)$.

By expanding brackets

$pqr = d$

$(4)^3 = 64$

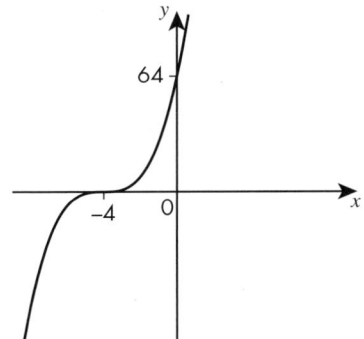

You may be given the cubic function in expanded form. In this case, you need to factorise it before you can sketch the curve. In **Chapter 2.3** you learned how to do this using the factor theorem.

EXAMPLE 21

Sketch the graph of the function $y = x^3 - 6x^2 + 11x - 6$.

Let $f(x) = x^3 - 6x^2 + 11x - 6$.
$f(1) = 1^3 - 6(1)^2 + 11(1) - 6 = 1 - 6 + 11 - 6 = 0$
So, as $f(1) = 0$, $(x - 1)$ is a factor of $x^3 - 6x^2 + 11x - 6$.
The constant term is -6 so try the other factors of -6: these are 2, 3, -1, -2, -3 and -6.
$f(2) = 2^3 - 6(2)^2 + 11(2) - 6 = 8 - 24 + 22 - 6 = 0$
So, as $f(2) = 0$, $(x - 2)$ is a factor of $x^3 - 6x^2 + 11x - 6$.
$f(3) = 3^3 - 6(3)^2 + 11(3) - 6 = 27 - 54 + 33 - 6 = 0$
So, as $f(3) = 0$, $(x - 3)$ is a factor of $x^3 - 6x^2 + 11x - 6$.
Therefore, $y = x^3 - 6x^2 + 11x - 6 = (x - 1)(x - 2)(x - 3)$.

Shape

$x \times x \times x = x^3$

$a = 1$. As $a > 0$, the curve has this shape: ∿

y-intercept

You can see from the original function that the constant term is -6. So the y-intercept is $(0, -6)$.

Roots

$(x - 1)(x - 2)(x - 3) = 0$

Roots are $x = 1$, $x = 2$, $x = 3$, so the x-intercepts are $(1, 0)$, $(2, 0)$ and $(3, 0)$.

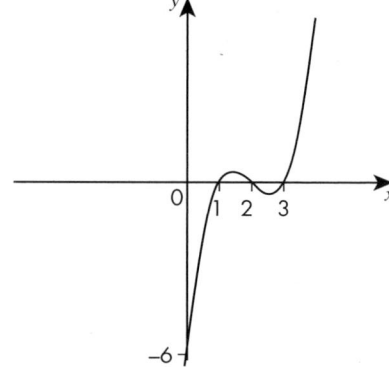

EXERCISE 3D

TARGETING DISTINCTION*

1. Sketch the graph of each function on separate axes. In each case, label the intercepts on the x-axis and y-axis.

   a) $y = (x - 3)(x - 1)(x + 5)$ **b)** $y = x(x - 1)(x + 3)$ **c)** $y = 3(x - 2)(x + 1)(x + 3)$

HINTS AND TIPS
In question **1c**, the number outside the brackets will not affect the roots, but will affect the y-intercept: $3pqr = 3((-2) \times 1 \times 3) = 3 \times (-6) = -18$

2. Sketch the graph of each function on separate axes. In each case, label the intercepts on the x-axis and the y-axis.

   a) $y = (x-3)^2(x-1)$ **b)** $y = x^2(x-1)$ **c)** $y = 2(x-2)(x+1)^2$

   d) $y = -(x-2)^3$

3. Factorise the right-hand side of each equation and then sketch the graph of each function on separate axes. In each case, label the intercepts on the x-axis and the y-axis.

   a) $y = x^3 - 6x^2 + 3x + 10$ **b)** $y = x^3 - 7x - 6$ **c)** $y = -x^3 - 3x^2 + 10x$

4. Determine the cubic function represented by the following graph. Give your answer in the form $y = ax^3 + bx^2 + cx + d$.

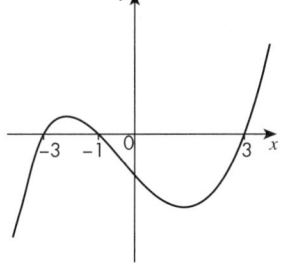

5. Determine the cubic function represented by the following graph. Give your answer in the form $y = ax^3 + bx^2 + cx + d$.

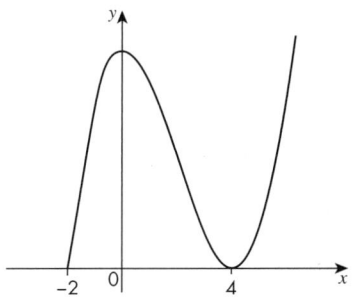

Quartic graphs

A quartic function contains an x^4 term as the highest power of x: for example, $y = 3x^4 + x^3 + x^2 - 3x + 2$, $y = 3x^4 + 2$, $y = 3x^4 - 3x^2 + 3x - 2$.

The general form of a quartic graph is $y = ax^4 + bx^3 + cx^2 + dx + e$, where $a > 0$.

You can sketch the graphs of quartic functions by considering where it meets or crosses the x-axis (points A, B, C and D on the diagrams) and its y-intercept (where it crosses the y-axis, that is, point E). A quartic graph usually (although not always) has three vertices or turning points (points F, G and H). As with sketching cubic graphs, you do not need to know the exact coordinates of these points.

Diagram 1 shows a positive quartic function ($a > 0$) and diagram 2 shows a negative quartic function ($a < 0$).

Notice that a quadratic graph (order 2) has one turning point. A cubic graph (order 3) has, at most, two turning points. A quartic (order 4) has up to three turning points. This pattern (order subtract one gives the maximum number of turning points) continues as the order of the polynomial continues to rise.

Diagram 1

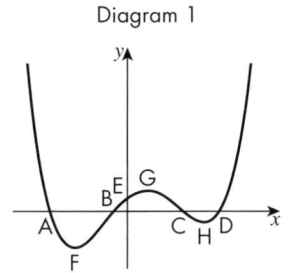

Diagram 2

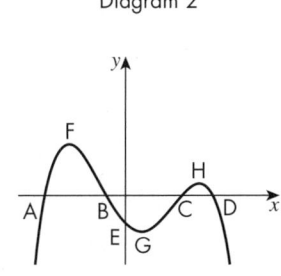

To sketch a quartic function, you can follow the same process as for sketching cubic functions. You can determine whether a quartic equation has repeated roots and whether a quartic graph has points of inflection using similar methods.

EXAMPLE 22

Sketch the curve of each of the following functions.

a) $y = (x - 3)(x - 1)(x + 2)(x + 5)$ b) $y = -x(x - 4)^2(x + 3)$
c) $y = (x - 2)^3(x + 2)$

> **HINTS AND TIPS**
>
> As this using a similar process to sketching cubic functions, you can simplify the working.

a)

Shape

$y = (x - 3)(x - 1)(x + 3)(x + 5)$
$a > 0$. Positive quartic curve.

Roots

$(x - 3)(x - 1)(x + 2)(x + 5) = 0$
Roots are $x = 3, 1, -2, -5$, so the x-intercepts are $(3, 0), (1, 0), (-2, 0)$ and $(-5, 0)$.

y-intercept

$(-3) \times (-1) \times 2 \times 5 = 30$
The y-intercept is $(0, 30)$.

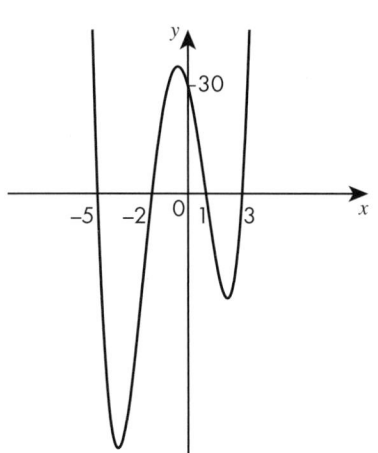

b)

Shape

$y = -x(x - 4)^2(x + 3)$
$a > 0$. Negative quartic curve.

Roots

$-x(x - 4)^2(x + 3) = 0$
Roots are $x = 0, 4, -3$
Repeated root at $x = 4$, so the graph cuts the x-axis at $(0, 0), (-3, 0)$ and it touches it at $(4, 0)$.

y-intercept is $(0, 0)$.

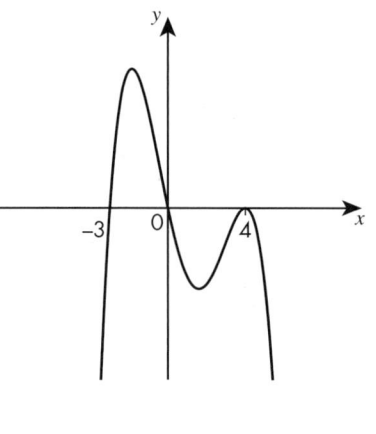

c)

Shape

$y = (x - 2)^3(x + 2)$
$a > 0$. Positive quartic curve.

Roots

$(x - 2)^3(x + 2) = 0$
$x = 2, -2$, so these are the x-intercepts
Point of inflection at $x = 2$.

y-intercept

$(-2)^3 \times 2 = -16$
The y-intercept is $(0, -16)$.

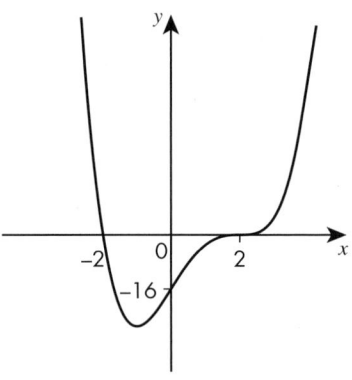

When a function has a quartic factor (for example, $x^4 + 12x^3 + 54x^2 + 108x + 81 = (x + 3)^4$, the graph just touches the x-axis in a similar way to a quadratic factor:

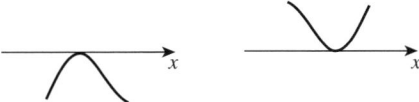

EXAMPLE 23

Sketch the curve of the function $y = (x - 2)^4$.

Shape

$a > 0$. Positive quartic curve.

Roots

$(x - 2)^4 = 0$

Repeated root at $x = 2$.

So the graph touches the x-axis at $(2, 0)$.

y-intercept

$(-2)^4 = 16$

So the y-intercept is $(0, 16)$.

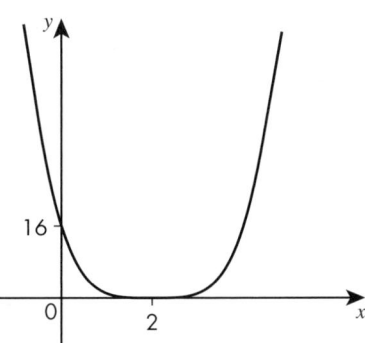

EXERCISE 3E

TARGETING DISTINCTION*

1. Sketch the graph of each function on separate axes. In each case, label the intercepts on the x-axis and the y-axis.
   a) $y = (x - 4)(x - 2)(x + 1)(x + 5)$
   b) $y = x(x - 2)(x + 1)(x + 3)$
   c) $y = 4(x - 3)(x - 2)(x + 1)(x + 3)$

2. Sketch the graph of each function on separate axes. In each case, label the intercepts on the x-axis and the y-axis.
   a) $y = (x - 2)^2(x - 1)(x + 2)$
   b) $y = x^2(x - 3)(x - 2)$
   c) $y = (x - 1)(x + 2)^3$
   d) $y = (x - 2)^4$

3.4 Trigonometric graphs

THIS SECTION WILL SHOW YOU HOW TO ...
- ✓ recognise the significant features of trigonometric graphs
- ✓ sketch the graphs of $y = \sin x$, $y = \cos x$ and $y = \tan x$

KEY WORDS
- ✓ asymptote
- ✓ cosine
- ✓ periodic
- ✓ sine
- ✓ tangent

$y = \sin x$, $0° \leqslant x \leqslant 360°$

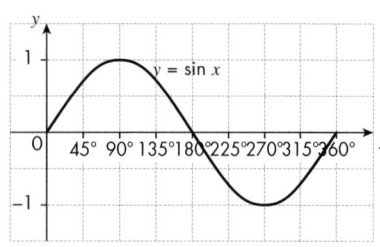

The **sine** graph is **periodic**, repeating every 360°.

$-1 \leqslant \sin x \leqslant 1$

When $90° < x < 180°$,
$\sin x = \sin (180° - x)$,
e.g. $\sin 153° = \sin 27°$

When $180° < x < 270°$,
$\sin x = -\sin (x - 180°)$,
e.g. $\sin 214° = -\sin 34°$
$= -0.559$

When $270° < x < 360°$,
$\sin x = -\sin (360° - x)$,
e.g. $\sin 287° = -\sin 73°$

$y = \cos x$, $0° \leqslant x \leqslant 360°$

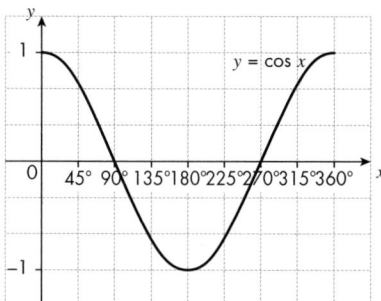

The **cosine** graph is periodic, repeating every 360°.

$-1 \leqslant \cos x \leqslant 1$

When $90° < x < 180°$,
$\cos x = -\cos (180° - x)$,
e.g. $\cos 161° = -\cos 19°$

When $180° < x < 270°$,
$\cos x = -\cos (x - 180°)$,
e.g. $\cos 245° = -\cos 65°$

When $270° < x < 360°$,
$\cos x = \cos (360° - x)$,
e.g. $\cos 310° = \cos 50°$

$y = \tan x$, $0° \leqslant x \leqslant 360°$

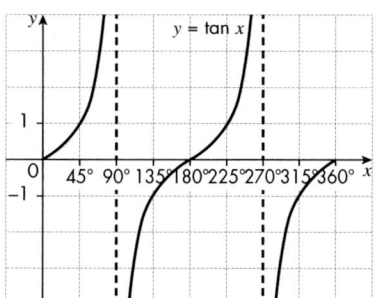

The **tangent** graph is periodic, repeating every 180°. It has **asymptotes** every $(180n + 90)°$ e.g. at $x = 90°$, $x = 270°$, $x = 450°$ etc.

$-\infty \leqslant \tan x \leqslant \infty$

When $90° < x < 180°$,
$\tan x = -\tan (180° - x)$,
e.g. $\tan 147° = -\tan 33°$

When $180° < x < 270°$,
$\tan x = \tan (x - 180°)$,
e.g. $\tan 197° = \tan 17°$

When $270° < x < 360°$,
$\tan x = -\tan (360° - x)$,
e.g. $\tan 302° = -\tan 58°$

HINTS AND TIPS

An asymptote is a straight line that a function approaches but never touches.

EXAMPLE 24

a) Using the sine graph, solve the equation $\sin x = 0.56$ for $0° \leqslant x \leqslant 360°$.

a) Find the first value using your calculator.
$\sin^{-1} 0.56 = 34.1°$
This is point A on the diagram.

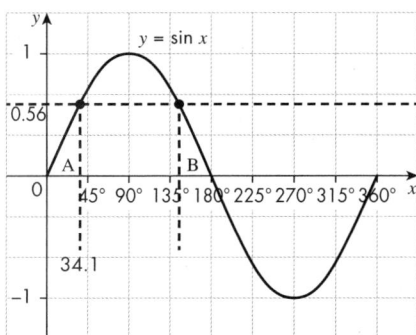

As $y = \sin x$ has a line of symmetry at $x = 90°$, the second angle can be marked on the graph. This is point B.

For $90° < x < 180°$, $\sin x = \sin(180° - x)$.
Therefore, $\sin 34.1° = \sin(180° - 34.1°)$
$= \sin 145.9°$
$x = 34.1°, 145.9°$

b) Using the cosine graph, solve the equation $\cos x = -0.285$ for $0° < x < 360°$.

b) $\cos^{-1}(-0.285) = 106.6°$

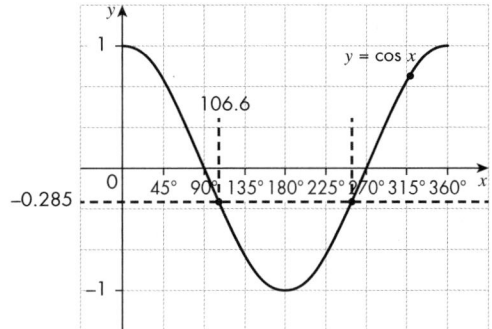

The symmetry of the curve, through $x = 180°$, shows us that the second value can be found in several ways:
$x = 360° - 106.6°$
$x = 253.4°$

$180° - 106.6° = 73.4°$
$x = 180° + 73.4°$
$x = 253.4°$

$106.6° - 90° = 16.6°$
$x = 270° - 16.6°$
$x = 253.4°$

So $x = 106.6°, 253.4°$

c) Using the tangent graph, solve the equation

c) $\tan x = -1.5$ for $0° \leqslant x \leqslant 360°$.
$\tan^{-1}(-1.5) = -56.3°$

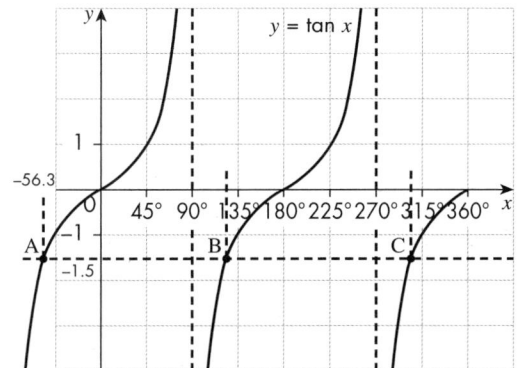

Therefore,
−56.3° + 180° = 123.7°
123.6° + 180° = 303.7°
x = 123.7°, 303.7°

As −56.3° is outside of the range that you want the answers to be in, you need to work out the values at point B and point C. As in the previous example, the distances between these points (A and B; B and C) is 180°.

EXERCISE 3F

TARGETING MERIT

1. State the two angles between 0° and 360° for each of these sine values.
   a) 0.6   b) 0.8   c) 0.75   d) −0.7   e) −0.25

2. Which of these values is the odd one out and why?
   sin 36°   sin 144°   sin 234°   sin 324°

3. The graph of sine x is cyclic, which means that it repeats indefinitely in each direction.
   a) Write down one value of x greater than 360° for which the sine has value 0.978 147 600 73.
   b) Write down one value of x less than 0° for which the sine has value 0.978 147 600 73.
   c) Describe any symmetries of the graph of $y = \sin x$.

4. State the two angles between 0° and 360° for each of these cosine values
   a) 0.6   b) 0.58   c) 0.458   d) −0.8   e) −0.25

5. Which of these values is the odd one out and why?
   cos 58°   cos 118°   cos 238°   cos 262°

6. The graph of cosine x is cyclic, which means that it repeats indefinitely in each direction.
   a) Write down one value of x greater than 360° for which the cosine value is −0.669 130 606 36.
   b) Write down one value of x less than 0° for which the cosine value is −0.669 130 606 36.
   c) Describe any symmetries of the graph of $y = \cos x$.

7. State the angles between 0° and 360° which have each of these tangent values.
   a) 0.258   b) 0.785   c) 1.19   d) −0.358   e) −0.634

8. Which of these values is the odd one out and why?
   tan 45°   tan 135°   tan 235°   tan 315°

9. The graph of tan x is cyclic, which means that it repeats indefinitely in each direction.
   a) Write down one value of x greater than 360° for which the tangent value is 2.144 506 920 51.
   b) Write down one value of x less than 0° for which the tangent value is 2.144 506 920 51.
   c) Describe any symmetries of the graph of $y = \tan x$.

TARGETING DISTINCTION

10. Write down the two possible values of x ($0° < x < 360°$) for each equation. Give your answers to 1 decimal place.

    a) $\sin x = 0.361$ b) $\sin x = -0.486$ c) $\cos x = 0.641$

11. Find two angles such that the sine of each is 0.5.

12. a) Given that $\sin 65° = 0.906$, find another angle between 0° and 360° that also has a sine of 0.906.
    b) Given that $\sin 213° = -0.545$, find another angle between 0° and 360° that also has a sine of −0.545.
    c) Given that $\cos 36° = 0.809$, find another angle between 0° and 360° that also has a cosine of 0.809.

13. a) Given that $\sin 30° = 0.5$, find two angles between 0° and 360° that have a sine of −0.5.
    b) Given that $\cos 45° = 0.707$, find two angles between 0° and 360° that have a cosine of −0.707.

14. $\cos 41° = 0.755$. What is $\cos 139°$?

TARGETING DISTINCTION*

15. Find the two solutions of each of these equations for $0° \leq x \leq 360°$.

    a) $\sin(x + 20°) = 0.5$ b) $\cos(5x) = 0.45$

16. a) Choose an acute angle a. Write down the values of:
       i) $\sin a$ ii) $\cos(90° - a)$.
    b) Repeat with another acute angle b.
    c) Write down a rule connecting the sine of an acute angle x and the cosine of the complementary angle (that is, when the angles add to 90°).

> **HINTS AND TIPS**
>
> This relationship between sine and cosine is why cosine is so named: 'complementary sine = cosine'.

17. Given that $\sin 26° = 0.438$

    a) write down an angle between 0° and 90° that has a cosine of 0.438
    b) find two angles between 0° and 360° that have a sine of −0.438
    c) find two angles between 0° and 360° that have a cosine of −0.438.

18. State whether each of the following rules is true or false.

    a) $\sin x = \sin(180° - x)$ b) $\cos x = \cos(360° - x)$
    c) $\sin x = -\sin(180° + x)$ d) $\cos(180° - x) = \cos(180° + x)$
    e) $\tan x = \tan(180° + x)$ f) $\tan(180° - x) = \tan(180° + x)$

Exam-style questions

1. The line l_1 passes through the points A(3, −5) and B(−2, 7).
   a) Find an equation for l_1 in the form $ax + by + c = 0$. [2 marks]
   b) The line l_2 is the perpendicular bisector of l_1. Find the equation of the line l_2. [3 marks]

   HINTS AND TIPS
   If the question does not dictate which form your answer should be in, give it in the form that is easiest to find. In most cases, this will be $y = mx + c$.

2. $\sin x = 0.454$ to 3 decimal places.
   One solution for x is 27°.
   a) Find a second solution between 0° and 360°. [1 mark]
   b) Hence, or otherwise, solve $\sin y = -0.454$, $0° \leq x \leq 360°$. [2 marks]

3. The graph of $y = ax^2 + bx + c$ has a maximum point at $\left(-\frac{2}{3}, -\frac{2}{3}\right)$. Given that the graph meets the y-axis at $(0, -2)$, find the values of a, b and c. [3 marks]

4. A ball is thrown into the air from the ground. The path of the ball can be modelled by the graph $h = 3t - \frac{1}{2}t^2$, where h is the height, in metres, of the ball above the ground and t is the time, in seconds, since the ball was thrown. After how many seconds does the ball hit the ground? [3 marks]

5. The diagram shows the graph of a quartic function, f(x). $(x + 4)^2$ is a factor of f(x).
   Write the function, f(x), in the form $y = (x + p)^2(x + q)(x + r)$, where p, q and r are real numbers.

   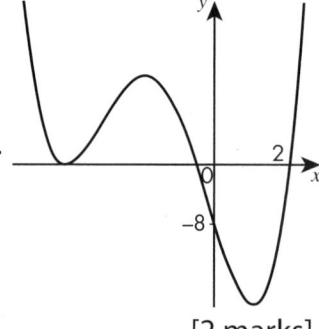

   [2 marks]

6. a) Sketch the graph of $y = \cos x$ for $-180° \leq x \leq 180°$. [2 marks]
   b) Using your graph, solve $\cos x = 0.85$ for $-180° \leq x \leq 180°$. [2 marks]

7. The table shows the cosines of some angles, correct to 2 decimal places.

X	10°	20°	70°	80°
cos x	0.98	0.94	0.34	0.17

   Use the values in the table to find:
   a) cos 100° [1 mark]
   b) cos 200° [1 mark]
   c) sin 10° [1 mark]

8. Find the equation of the line that is parallel to $2x + 3y = 6$ and passes through the point (4, 4). [3 marks]

9. The equation $(q + 1)x^2 + 2\sqrt{2}qx + (2q - 1) = 0$ has two equal roots. Show that q is an integer. [4 marks]

10. a) Sketch the graph of $y = (x + 3)^2(x - 1)$. [2 marks]

   b) f(x) is a quadratic function. Given that af(x) meets the x- and y-axes at the same points as $y = (x + 3)^2(x - 1)$, find the value of a. [3 marks]

11. A circle with centre $(-4, 2)$ has a tangent at the point $(-1, 6)$. The tangent meets the y-axis at point A and the x-axis at point B.

    a) Find the coordinates of points A and the x-axis at point B. [4 marks]

    b) Find the area of the triangle formed by joining points A and B and the origin, O. [3 marks]

4 More graphs

4.1 Translating and reflecting graphs

THIS SECTION WILL SHOW YOU HOW TO ...
- ✓ Translate graphs of functions
- ✓ Reflect graphs of functions
- ✓ Translate and reflect trigonometric graphs

KEY WORDS
- ✓ reflection
- ✓ transformation
- ✓ translation

Translating graphs
Consider the following graphs.

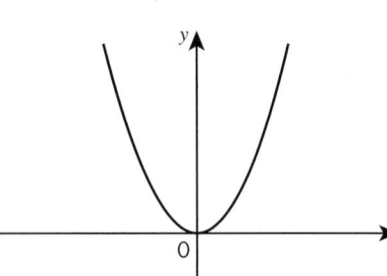
$y = x^2$

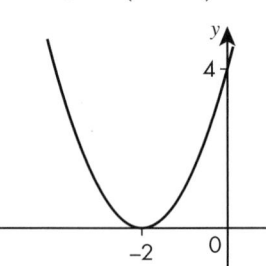
$y = (x + 2)^2$

This is a **translation** of $y = x^2$ by the vector $\begin{pmatrix} -2 \\ 0 \end{pmatrix}$.

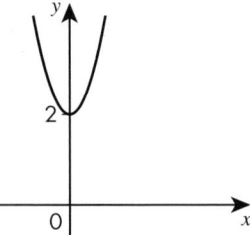
$y = x^2 + 2$

This is a translation of $y = x^2$ by the vector $\begin{pmatrix} 0 \\ 2 \end{pmatrix}$.

All points on the graph undergo the same translation, so the shape of the graph is unchanged. Only its position is affected by the **transformation**.

In general: $y = f(x + a)$ translates $y = f(x)$ by the vector $\begin{pmatrix} -a \\ 0 \end{pmatrix}$

When a constant is added or subtracted 'inside' the function, it results in a horizontal translation.

HINTS AND TIPS

You can think of a horizontal translation as moving in the opposite direction to what you might expect. For example, you may have expected $y = (x + 2)^2$ to translate $y = x^2$ by 2 units to the right, as 2 has been added to x inside the bracket. However, if you think back to **Chapter 3**, you can see why this is not the case. A graph meets the x-axis where the function is equal to zero, so, for the translation $y = (x + 2)^2$,

$(x + 2)^2 = 0$
$x = -2$

Therefore, the minimum point has moved left from (0, 0) to (−2, 0).

$y = f(x) + a$ **translates** $y = f(x)$ **by the vector** $\begin{pmatrix} 0 \\ a \end{pmatrix}$

When a constant is added or subtracted 'outside' the function, it results in a vertical translation.

> **HINTS AND TIPS**
>
> You can think of a vertical translation as moving in the direction that you would expect. For example, $y = x^2 + 2$ translates the graph of $y = x^2$ by 2 units upwards (2 units in the positive y-direction).

These transformations can be combined.

EXAMPLE 1

The sketch shows the graph of $f(x) = (x + 2)(4 - x)$. Sketch the following functions, labelling the maximum point on each graph.

a) $f(x) + 3$
b) $f(x - 1)$
c) $f(x + 2) - 1$

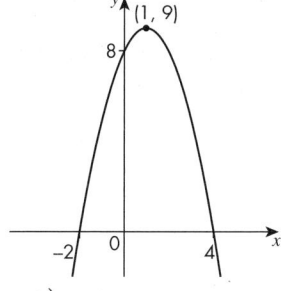

a) $f(x) + 3$

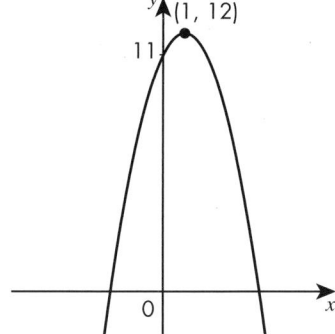

The graph of $f(x)$ is translated by the vector $\begin{pmatrix} 0 \\ 3 \end{pmatrix}$. The y-intercept and the y-coordinate of the maximum point increase by 3.

b) $f(x - 1)$

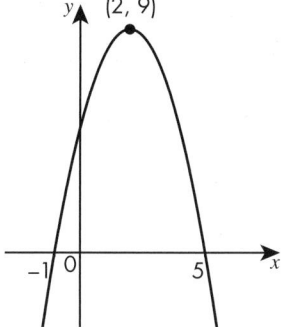

The graph of $f(x)$ is translated by the vector $\begin{pmatrix} 1 \\ 0 \end{pmatrix}$. The x-intercepts and the x-coordinate of the maximum point increase by 1.

c) $f(x + 2) - 1$

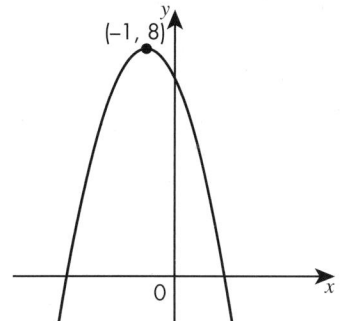

The graph of $f(x)$ is translated by the vector $\begin{pmatrix} -2 \\ -1 \end{pmatrix}$. The maximum point has moved two units left and one unit down.

> **HINTS AND TIPS**
>
> This is perhaps counter-intuitive, but remember that when the addition or subtraction is happening inside the function bracket, the translation acts in the opposite direction to what you might expect.

Trigonometric examples work in the same way.

EXAMPLE 2

The diagram shows the graph of $f(x) = \sin x$ for $0° \leqslant x \leqslant 360°$. The x-intercepts, y-intercept and vertices are labelled.

Sketch:

a) $f(x + 90°)$

b) $f(x) - 1$

c) $f(x - 90°) + 2$

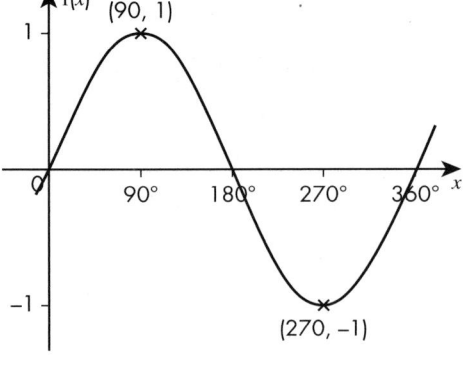

$f(x + 90°) = \sin(x + 90°)$ $f(x) - 1 = \sin(x) - 1$ $f(x - 90°) + 2 = \sin(x - 90°) + 2$

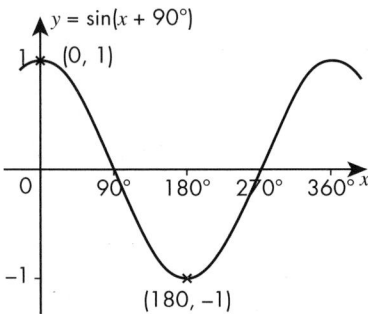 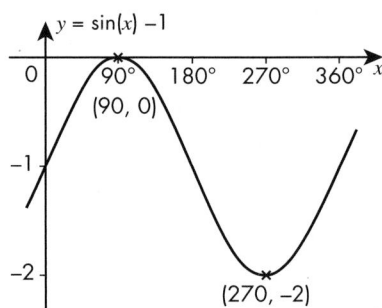 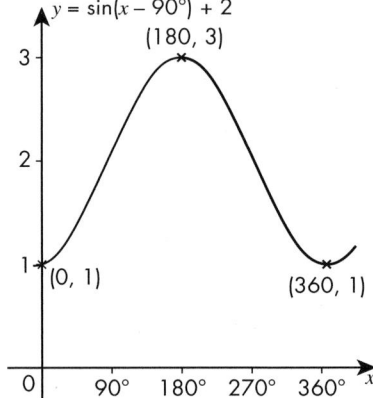

The graph of $f(x)$ is translated by vector $\begin{pmatrix} -90 \\ 0 \end{pmatrix}$.

This graph should look familiar – it is the cosine graph: $\sin(x + 90°) = \cos x$

The graph of $f(x)$ is translated by vector $\begin{pmatrix} 0 \\ -1 \end{pmatrix}$.

The graph of $f(x)$ is translated by vector $\begin{pmatrix} 90 \\ 2 \end{pmatrix}$.

Reflecting graphs

Consider the following functions and their graphs.

$f(x) = (x + 3)^3$ $f(-x) = ((-x) + 3)^3 = (3 - x)^3$ $-f(x) = -(x + 3)^3$

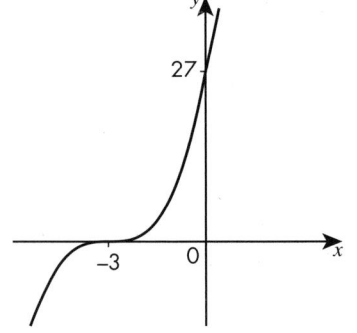 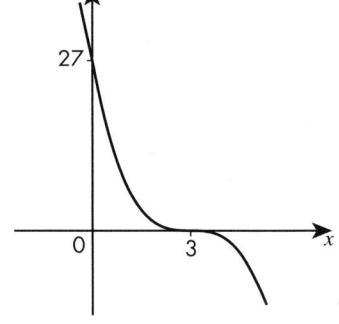 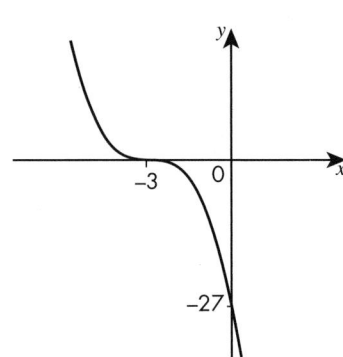

The graph of $y = f(-x)$ is a **reflection** of the graph of $f(x)$ in the y-axis.

The graph of $-f(x)$ is a reflection of the graph of $y = f(x)$ in the x-axis.

HINTS AND TIPS

The y-intercept remains the same, and the x-intercept is reflected (changes sign).

HINTS AND TIPS

The x-intercept remains the same, and the y-intercept is reflected (changes sign).

In general:
$y = f(-x)$ **reflects** $y = f(x)$ **in the y-axis.**
$y = -f(x)$ **reflects** $y = f(x)$ **in the x-axis.**

EXAMPLE 3

a) Sketch the quartic function $f(x) = x(x+3)(x-2)^2$.
b) Sketch: **i)** $-f(x)$ **ii)** $f(-x)$

a) $f(x) = x(x+3)(x-2)^2$

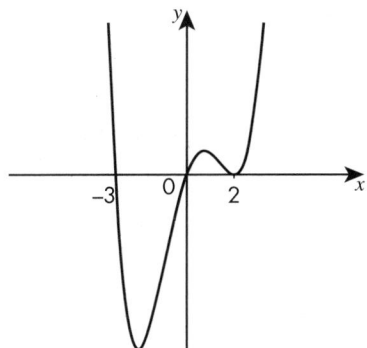

b) i)

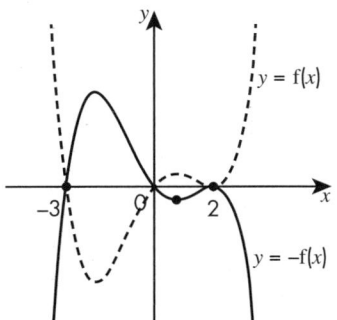

ii)

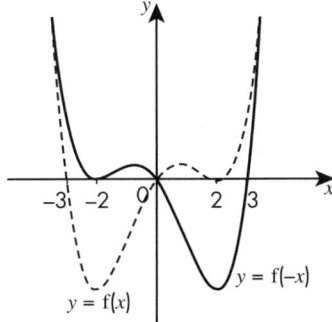

You can also reflect trigonometric graphs.

EXAMPLE 4

The diagram shows the graph of $f(x) = \cos x$ for $-180° \leqslant x \leqslant 180°$.

The x-intercepts, y-intercept and vertices are labelled.
Sketch:
a) $-f(x)$
b) $f(-x)$

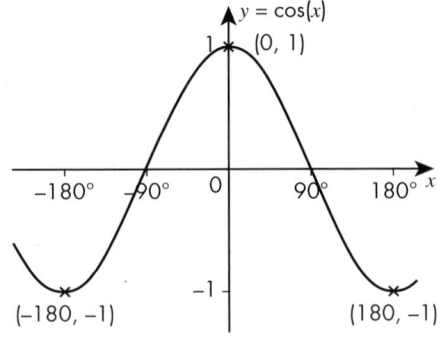

a)
$-f(x) = -\cos x$

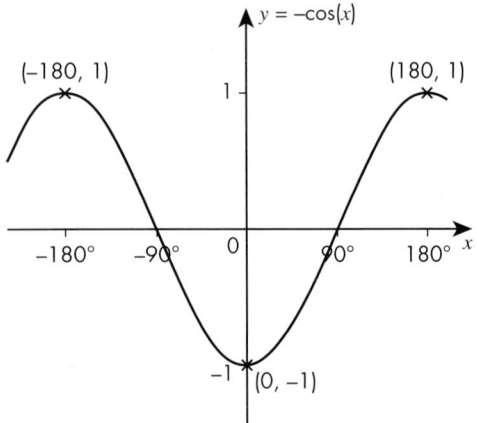

The curve has been reflected in the x-axis.

b)
$f(-x) = \cos(-x)$

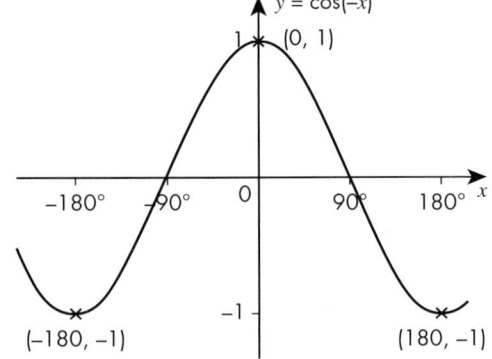

The curve has been reflected in the y-axis, but as the y-axis is a line of symmetry for the cosine graph, the diagram is unchanged.

EXERCISE 4A

TARGETING MERIT

1. The diagram shows the graph of $y = f(x)$.
   The vertices $(-2, 3)$ and $(0, -1)$ are shown.
   Sketch the following functions, labelling the coordinates of the vertices.

   a) $y = f(x - 2)$
   b) $y = f(x) + 1$
   c) $y = f(x) - 3$
   d) $y = f(x + 1)$
   e) $y = f(-x)$
   f) $y = -f(x)$

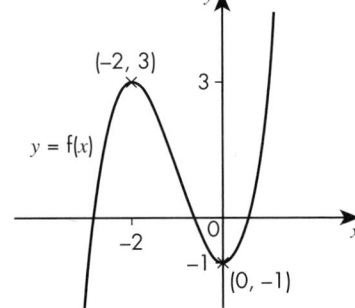

2. The points A(−1, 2), B(0, 3) and C(4, 7) lie on the graph of $y = g(x)$.

   State the coordinates of A, B and C after the following transformations.

   a) $y = f(x + 3)$ b) $y = f(x) − 2$ c) $y = f(x) + 7$

   d) $y = f(x − 4)$ e) $y = -f(x)$ f) $y = f(-x)$

TARGETING DISTINCTION

3. a) Sketch the curve $y = f(x)$ where $f(x) = x^2 − 7x + 12$. Label the x-intercepts, y-intercept and vertex of each graph.

   b) Sketch the following transformations, labelling, where possible, the x-intercepts, y-intercept and vertex of each graph.

   i) $y = f(x + 3)$ ii) $y = f(x) + 3$ iii) $y = -f(x)$ iv) $y = f(-x)$

4. The graph of $y = f(x)$ is translated. The point Q(−1.5, 4) lies on $y = f(x)$. On the transformed curve, point Q has coordinates (4, −1.5). Write the equation of the translated curve in the form $y = f(x + a) + b$.

TARGETING DISTINCTION*

5. $f(x) = \tan x$

   The diagram shows $y = f(x)$ for $0° \leq x \leq 360°$.

   Sketch the following transformations labelling, where possible, the x-intercepts, y-intercept and asymptotes of each graph.

   a) $y = f(x − 90°)$

   b) $y = f(x) + 1$

   c) $y = f(x + 90°) − 1$

   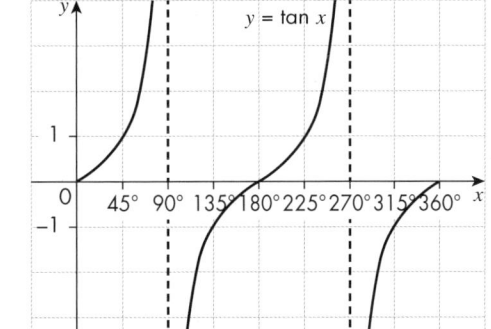

6. $f(x) = (x − 2)^2$

   Transform the graph of $y = f(x)$, by applying the following translations and reflections. Label the vertex of the graph in each diagram.

   a) $y = -f(x + 2)$ b) $y = f(x − 3) + 1$ c) $y = f(-x) + 4$

   d) $y = -f(x + 1) − 3$ e) $y = f(-x + 1) + 2$ f) $y = -f(-x)$

   HINTS AND TIPS

   Asymptotes are affected by transformations in the same way as points on a graph.

   HINTS AND TIPS

   When applying multiple transformations to a curve, follow this order: 1) horizontal translation; 2) reflection in the y-axis; 3) reflection in the x-axis; 4) vertical translation.

7. $f(x) = \sin x$

The diagram shows $y = f(x)$ for $0° \leq x \leq 360°$.

Sketch the following transformations labelling, where possible, the x-intercepts, y-intercept and vertices of each graph.

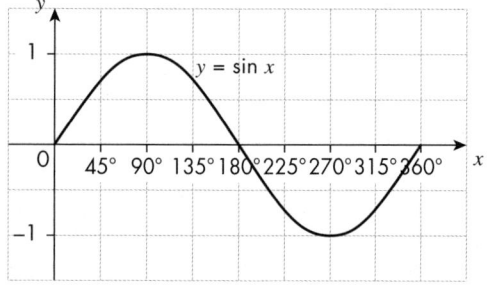

a) $y = -f(x + 90°)$ b) $y = f(-x) - 1$

c) $y = -f(-x)$

8. The points A$(-1, 2)$ and B$(0, 3)$ lie on the graph of $y = f(x)$.

State the coordinates of A and B after the following transformations.

a) $-y = f(x) + 1$ b) $y - 2 = f(x) + 2$ c) $3 - y = f(x)$

4.2 Stretching graphs

THIS SECTION WILL SHOW YOU HOW TO ...

✓ Stretch graphs of functions, including trigonometric graphs

KEY WORDS

✓ stretch

Stretching graphs

Consider the following functions and their graphs.

$f(x) = (x + 3)^3$ $f(2x) = (2x + 3)^3$ $2f(x) = 2(x + 3)^3$

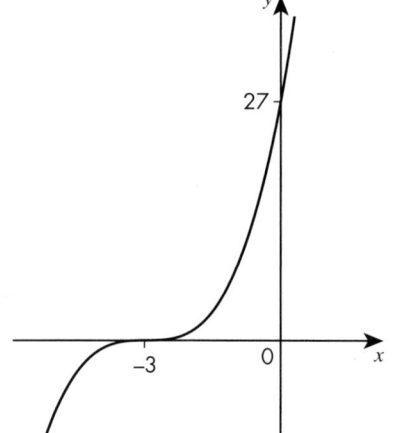   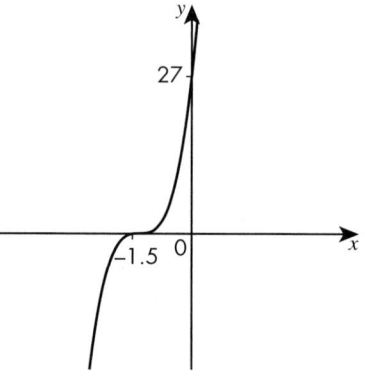   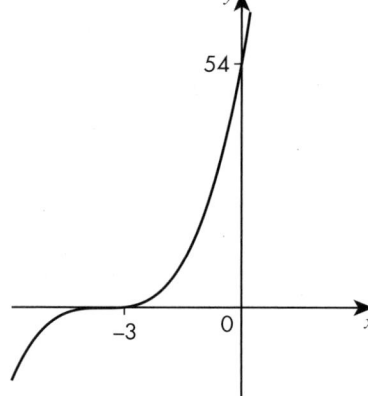

The y-intercept is unaffected, but the x-intercept is divided by 2 (or multiplied by $\frac{1}{2}$).

The x-intercept is unaffected, but the y-intercept is multiplied by 2.

The transformations f(2x) and 2f(x) represent stretches.
In general:

y = f(ax) stretches y = f(x) in the x-direction (horizontally) by a factor of $\frac{1}{a}$.

This means that x-coordinates are affected but y-coordinates are not.

y = af(x) stretches y = f(x) in the y-direction (vertically) by a factor of a.

This means that y-coordinates are affected but x-coordinates are not.

> **HINTS AND TIPS**
>
> As for translations, you can think of stretches in the x direction as being counter intuitive. For y = f(2x), you might expect the x-coordinates to double, but instead, they are halved.

> **HINTS AND TIPS**
>
> When the stretch is given 'inside' the function, for example, (f(3x)), then the stretch is horizontal. If the stretch is 'outside' the function, for example, (4f(x)), then the stretch is vertical.

EXAMPLE 5

The sketch shows a quartic function y = g(x), with the coordinates of the vertices labelled.

Sketch:
a) 2g(x)   b) g(3x)   c) $g\left(\frac{1}{2}x\right)$

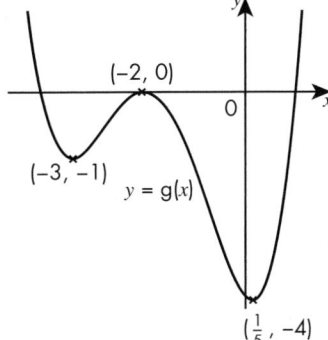

a) 2g(x)   b) g(3x)   c) $g\left(\frac{1}{2}x\right)$

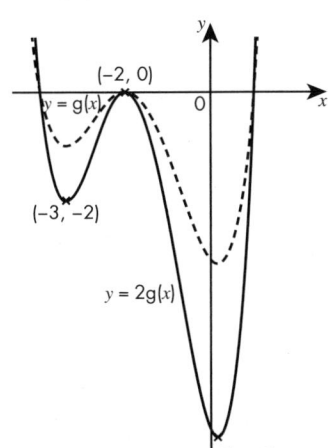 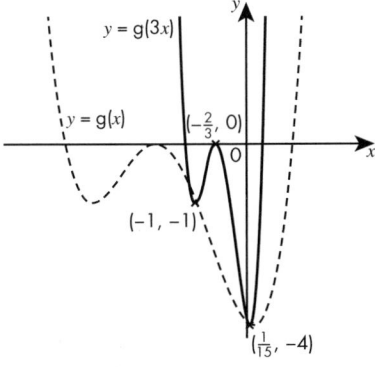 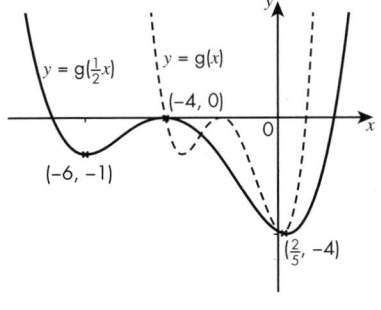

You can also stretch trigonometric graphs.

EXAMPLE 6

The diagram shows the graph of $f(x) = \sin x$ for $0° \leqslant x \leqslant 360°$.
The x-intercepts, y-intercept and vertices are labelled.

Sketch:
a) $2f(x)$
b) $f(3x)$

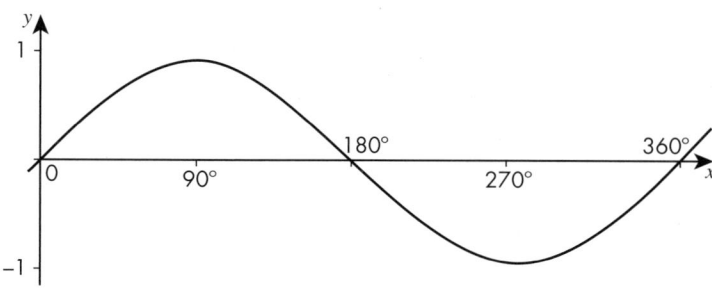

a) $2f(x) = 2\sin x$

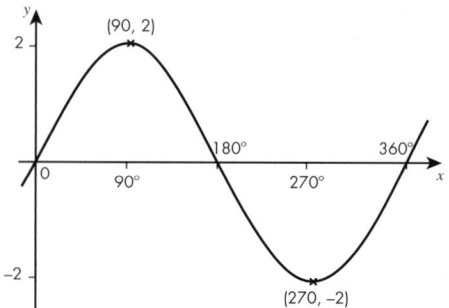

b) $f(3x) = \sin(3x)$

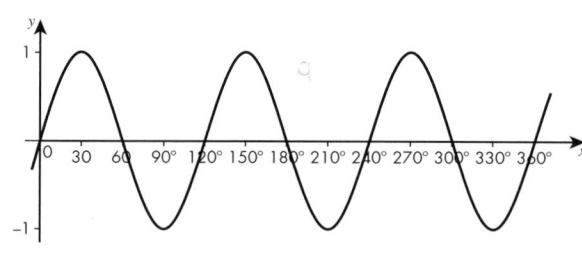

EXERCISE 4B

TARGETING DISTINCTION*

1. The diagram shows the graph of $y = f(x)$. The vertices $(-2, 3)$ and $(0, -1)$ are shown.

   Sketch the following functions, labelling the coordinates of the vertices.

   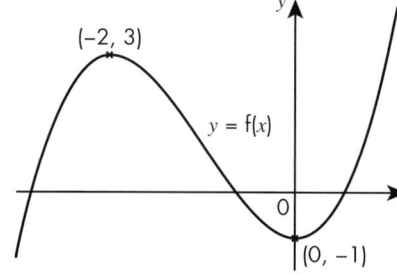

   a) $y = 2f(x)$
   b) $y = f(2x)$
   c) $y = \tfrac{1}{2}f(x)$
   d) $y = f\left(\tfrac{1}{2}x\right)$

2. The points $A(-1, 2)$, $B(0, 3)$ and $C(4, 7)$ lie on the graph of $y = g(x)$.
   State the coordinates of A, B and C after the following transformations.
   a) $y = 3f(x)$
   b) $y = -2f(x)$
   c) $y = f(5x)$
   d) $y = f\left(\tfrac{1}{4}x\right)$
   e) $y = f(-2x)$
   f) $y = \tfrac{1}{3}f(x)$

3. **a)** Sketch the graph of $y = f(x)$, where $f(x) = -x^2 - 3x + 10$. Label the x-intercepts, y-intercept and vertex.
   b) Sketch the following stretches, labelling the x-intercepts, y-intercept and vertex of each graph.

   i) $y = 3f(x)$ **ii)** $y = \frac{1}{2}f(x)$ **iii)** $y = f(2x)$

   iv) $y = f\left(\frac{1}{3}x\right)$ **v)** $y = -2f(x)$ **vi)** $y = f(-3x)$

4. Sketch, on the same axes, $y = \cos x$, $y = 2\cos x$ and $y = \cos 2x$, for $0° \leq x \leq 360°$.
   Label the x-intercepts, y-intercept and vertices of each graph.

5. The graph of $y = f(x)$, where $f(x) = \tan x$, passes through the points $(0, 0)$, $(45, 1)$ and $(135, -1)$.
   State the coordinates of these points after each of the following transformations.
   a) $y = 3f(x)$ **b)** $y = f\left(\frac{1}{2}x\right)$ **c)** $y = 2f(2x)$

6. **a)** Sketch the graph of the function $y = f(x)$, where $f(x) = x^3 + 4x^2 + 4x$.
   b) Sketch the graph of $y = f(4x)$.

7. Sketch the curve $2y = \sin x$.

8. The point $P(-2, 7)$ lies on the graph $y = f(x)$. State the coordinates of P on the graph of $y = -2f(-3x) + 1$.

4.3 Circles

THIS SECTION WILL SHOW YOU HOW TO ...
✓ Recognise the equation of a circle, centre (0, 0) or (a, b), radius r
✓ Understand that the circle $(x - a)^2 + (y - b)^2 = r^2$ is a translation of the circle $x^2 + y^2 = r^2$ by the vector $\begin{pmatrix} a \\ b \end{pmatrix}$
✓ Sketch circles from their equations

The equation of a circle.

The general equation of a circle with centre (0, 0) and radius r is $x^2 + y^2 = r^2$.

This can be shown using Pythagoras' theorem.

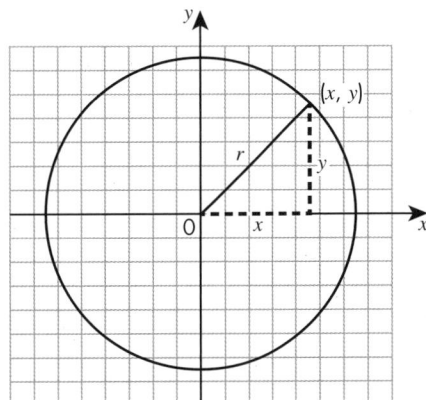

Pythagoras' theorem states that, for a right-angled triangle,
$a^2 + b^2 = c^2$, where c is the hypotenuse of the triangle.
This leads to $x^2 + y^2 = r^2$.

For example, the equation of a circle with centre (0, 0) and radius 5 is $x^2 + y^2 = 25$.

The point (3, 4) lies on the circle as the values of x and y satisfy the equation of the circle, $3^2 + 4^2 = 25$.

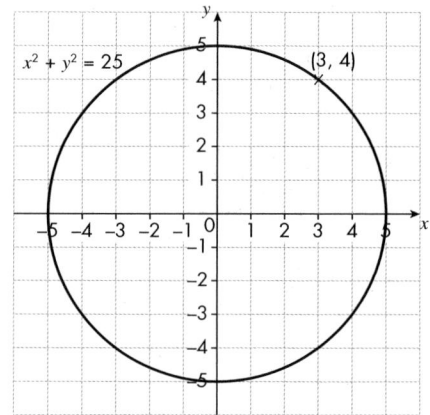

The general equation of a circle with centre (a, b) and radius r is $(x - a)^2 + (y - b)^2 = r^2$.

This represents a translation of the circle with centre (0, 0) by vector $\begin{pmatrix} a \\ b \end{pmatrix}$.

For example, the equation of a circle with centre (2, 6) and radius 13 is $(x - 2)^2 + (y - 6)^2 = 169$.

The point (−3, 18) lies on the circle as the values of x and y satisfy the equation of the circle:

$(-3 - 2)^2 + (18 - 6)^2 = 25 + 144 = 169$

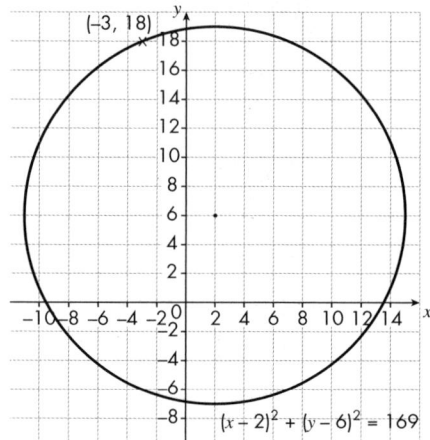

EXAMPLE 7

a) Write down the equation of a circle with centre (0, 0) and radius 3.
b) Write down the equation of a circle with centre (1, −4) and radius 7.

a) The equation is $x^2 + y^2 = 3^2$ or $x^2 + y^2 = 9$.
b) The equation is $(x - 1)^2 + (y + 4)^2 = 49$.

EXAMPLE 8

a) Show that the point (5, −2) lies on the circle with centre (0, 0) and radius $\sqrt{29}$.
b) Show that the point (−3, 2) lies on the circle with centre (5, 1) and radius $\sqrt{65}$.

a)

The equation of a circle, centre (0, 0) and radius $\sqrt{29}$, is $x^2 + y^2 = 29$.

At the point (5, −2), $x = 5$ and $y = -2$.

$5^2 + (-2)^2 = 25 + 4 = 29$

So, the point (5, −2) lies on the circle.

b)

The equation of a circle centre (5, 1) and radius $\sqrt{65}$, is $(x - 5)^2 + (y - 1)^2 = 65$.

At the point (−3, 2), $x = -3$ and $y = 2$.

$(-3 - 5)^2 + (2 - 1)^2 = 64 + 1 = 65$

So (−3, 2) lies on the circle.

EXAMPLE 9

The points P(6, 5) and Q(−2, 4) lie on the circumference of a circle. PQ is the diameter.

a) Find the length of:
  i) the diameter
  ii) the radius.
b) Find the coordinates of the centre of the circle.
c) Find the equation of the circle.

> **HINTS AND TIPS**
>
> To find the length of the diameter, use Pythagoras' theorem to find the distance between two points:
> $$l = \sqrt{(y_2 - y_1)^2 + (x_2 - x_1)^2}$$

a) i) Length of diameter
$= \sqrt{(4-5)^2 + (-2-6)^2}$
$= \sqrt{65}$

  ii) Length of radius
$= \sqrt{65} \div 2$
$= \frac{\sqrt{65}}{2}$

b) The centre is the midpoint of the diameter. The coordinates are
$\left(\frac{6+(-2)}{2}, \frac{5+4}{2}\right) = \left(\frac{4}{2}, \frac{9}{2}\right)$
$= \left(2, \frac{9}{2}\right)$

c) The equation is
$(x-2)^2 + \left(y - \frac{9}{2}\right)^2 = \left(\frac{\sqrt{65}}{2}\right)^2$
$(x-2)^2 + \left(y - \frac{9}{2}\right)^2 = \frac{65}{4}$

The equation of a circle may be given in expanded form, for example,
$$(x-1)^2 + (y+4)^2 = 49$$
$$x^2 - 2x + 1 + y^2 + 8y + 16 = 49$$
$$x^2 + y^2 - 2x + 8y - 32 = 0$$

To return to the form $(x-a)^2 + (y-b)^2 = r^2$, from which you can identify the centre and radius of the circle, you can complete the square.

EXAMPLE 10

A circle has equation $x^2 + y^2 - 4x - 10y - 37 = 0$. Find the centre and radius of the circle.

Rearrange so that the x and y terms are together:
$x^2 - 4x + y^2 - 10y - 37 = 0$

Complete the square on $x^2 - 4x$ and $y^2 - 10x$ separately:
$(x-2)^2 - 4 + (y-5)^2 - 25 - 37 = 0$

Simplify and rearrange to $(x-a)^2 + (y-b)^2 = r^2$:
$(x-2)^2 + (y-5)^2 = 66$

State the centre and radius:
Centre has coordinates (2, 5)
Radius has length $\sqrt{66}$

EXERCISE 4C

TARGETING PASS

1. **a)** Write down the equation of each circle.
    b) Sketch each circle for **i)** to **iii)**.

    i) Centre (0, 0), radius 7
    ii) Centre (0, 0), radius 9
    iii) Centre (3, 3), radius $\sqrt{5}$
    iv) Centre (a, b), radius $2\sqrt{3}$

2. Write down the exact value of the radius for each of these circles.

    a) $x^2 + y^2 = 100$
    b) $x^2 + y^2 = 225$
    c) $(x - 2)^2 + (y + 3)^2 = 20$
    d) $(x + p)^2 + (y - q)^2 = 24$

    > **HINTS AND TIPS**
    > To find the radius of the circle, first find the diameter using Pythagoras' theorem, as shown in **Example 9**.

3. The line segment AB is the diameter of a circle.

    A is the point (1, −4) and B is the point (5, 2).
    Work out the equation of the circle.

4. The circle $x^2 + y^2 = 16$ is translated to the circle $(x - 4)^2 + (y + 7)^2 = 16$.

    Write down the vector for this translation.

5. The circle $x^2 + y^2 = 25$ is translated by the vector $\begin{pmatrix} -1 \\ 2 \end{pmatrix}$.

    Write down the equation of the circle after the translation.

TARGETING MERIT

6. Point A(4, 2) lies on the circumference of the circle with equation $x^2 + y^2 = 20$.
    a) Find the gradient of the line segment joining A to the centre of the circle.
    b) Find the gradient of the tangent to the circle at A.
    c) Find the equation of the tangent to the circle at A in the form $y = mx + c$.

    > **HINTS AND TIPS**
    > As covered in **Chapter 3**, the tangent and radius of a circle are perpendicular. This means that their gradients are negative reciprocals of each other.

7. Show that the tangent to the circle $x^2 + y^2 = 73$ at the point (−8, 3) is given by the equation $3y = 8x + 73$.

8. In each part, find the equation of the tangent in the form $y = mx + c$.

    a) Circle $x^2 + y^2 = 34$, tangent at (3, −5)
    b) Circle $x^2 + y^2 = 40$, tangent at (−2, −6)
    c) Circle $x^2 + y^2 = a^2$, tangent at (p, q)

9. A circle with centre at the origin has a tangent with equation $y = -\frac{1}{3}x + c$ at the point (3, 9).

    a) Find the value of c.
    b) Find the equation of the circle.

10. Point P(−2, −3) lies on a circle. The equation of the circle is $(x + 4)^2 + (y - 1)^2 = r^2$. Find the value of r.

11. P is a point (−1, 4) on the circle $x^2 + y^2 = 17$. Find the equation of the tangent to the circle at P.

TARGETING DISTINCTION

12. P is a point (5, −2) on the circle $(x-3)^2 + (y+1)^2 = 5$. Find the equation of the tangent to the circle at P.

13. Point A (2, 1) lies on a circle. The centre of the circle is (3, −2).

Find the equation of the circle.

TARGETING DISTINCTION*

14. Find the centre and radius of the following circles.

   a) $x^2 + y^2 - 8x + 6y + 16 = 0$ **b)** $x^2 + y^2 + 4x - 10y + 28 = 0$

   c) $x^2 + y^2 - 6x - 6y - 14 = 0$ **d)** $x^2 + y^2 + 10x - y - \frac{99}{4} = 0$

15. The circle $x^2 + y^2 + 2mx - 5ny + 17 = 0$ has a centre (−3, 4) and a radius of r. Find the values of n, m and r. Give your answers in exact form.

4.4 Exponential and reciprocal graphs

THIS SECTION WILL SHOW YOU HOW TO ...
✓ Plot and sketch exponential and reciprocal graphs
✓ Understand transformations on exponential and reciprocal functions

KEY WORDS
✓ asymptote   ✓ exponential function   ✓ reciprocal function

Exponential graphs

Equations of the form $y = ab^x$ and $y = ab^{-x}$, where a and b are rational numbers, are **exponential functions**.

The graph of $y = ab^x$ with $b > 1$ has these properties.

- It intersects the y-axis at $(0, a)$ because $b^0 = 1$.
- The value of y increases steeply as x increases.
- As x takes on increasingly large negative values, y gets closer to zero but never reaches zero. The negative x-axis is an **asymptote** to the graph.

The graph of $y = ab^x$ with $0 < b < 1$ has these properties.

- It intersects the y-axis at $(0, a)$ because $b^0 = 1$.
- The value of y decreases steeply as x increases.
- As x takes on increasingly large positive values, y gets closer to zero but never reaches zero. Therefore, the positive x-axis is an asymptote to the graph.

The graph of $y = ab^{-x}$ is a reflection of $y = ab^x$ in the y-axis.

HINTS AND TIPS

Remember from the work in **Section 4.1** on reflecting functions that the graphs of $f(x)$ and $f(-x)$ are reflections of each other in the y-axis.

Consider a numerical example to show this:

$y = a \times 2^{-x} = a \times (2^{-1})^x = a \times \left(\frac{1}{2}\right)^x$

Therefore, $y = ab^{-x} = a \times \left(\frac{1}{b}\right)^x$

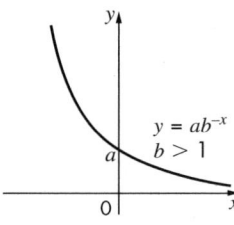

EXAMPLE 11

a) Complete the table of values for $y = 2^x$ for $-5 \leqslant x \leqslant 3$. Round each value to 2 decimal places.

x	−5	−4	−3	−2	−1	0	1	2	3
$y = 2^x$	0.03	0.06	0.13			1	2	4	

b) Plot the graph of $y = 2^x$ for $-5 \leqslant x \leqslant 3$.
c) Sketch the graph of $y = 2^x$.

a)

x	−5	−4	−3	−2	−1	0	1	2	3
$y = 2^x$	0.03	0.06	0.13	0.25	0.5	1	2	4	8

b)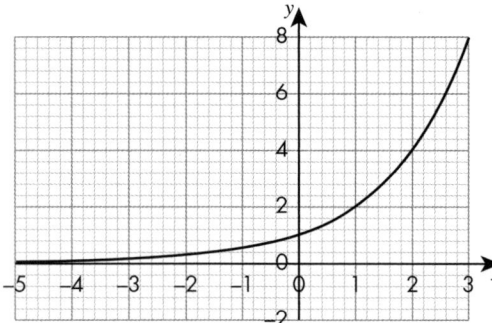

c) You know:
- the shape of an exponential function
- that it crosses the y-axis at $(0,1)$
- that it has an asymptote at $y = 0$.

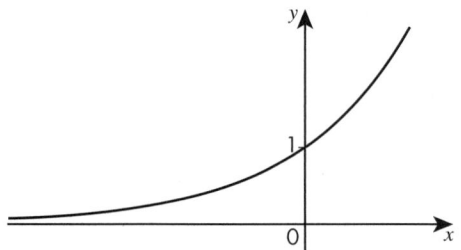

Reciprocal graphs

Equations of the form $y = \frac{a}{x}$ and $y = \frac{a}{x^2}$, where a is a real constant, are **reciprocal functions**.

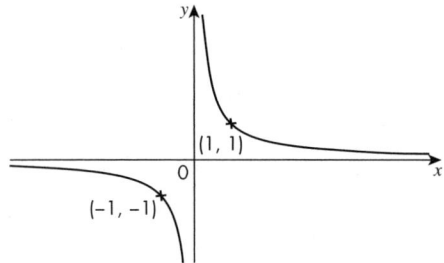
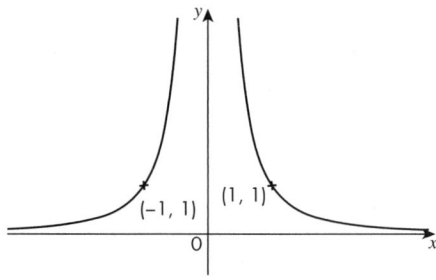

The graph of $y = \frac{1}{x}$ has these properties:
- It has asymptotes at $x = 0$, $y = 0$.
- The points $(1, 1)$ and $(-1, -1)$ lie on the curve. This can be generalised: for the graph $y = \frac{a}{x}$, the points $(1, a)$ and $(-1, -a)$ lie on the curve, because $\frac{a}{(1)} = a$ and $\frac{a}{(-1)} = -a$.

The graph of $y = \frac{1}{x^2}$ has these properties:
- It has asymptotes at $x = 0$, $y = 0$.
- The points $(1, 1)$ and $(-1, 1)$ lie on the curve. This can be generalised: for the graph $y = \frac{a}{x^2}$, the points $(1, a)$ and $(-1, a)$ lie on the curve, because $\frac{a}{(1)^2} = a$ and $\frac{a}{(-1)^2} = a$.

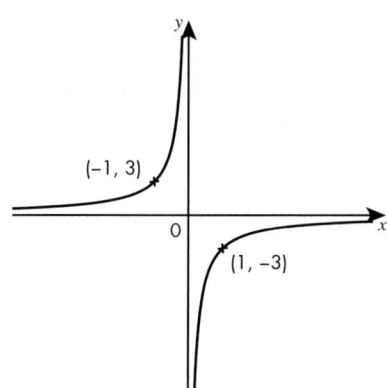

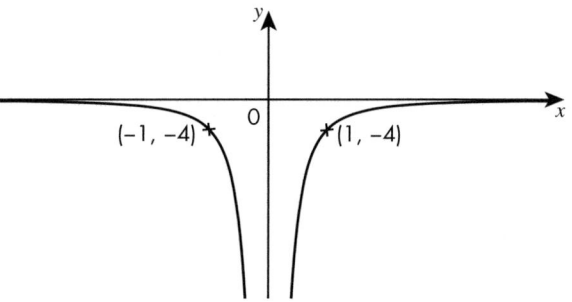

The graph of $y = -\frac{3}{x}$ has these properties:
- It has asymptotes at $x = 0$, $y = 0$.
- The points $(1, -3)$ and $(-1, 3)$ lie on the curve. This can be generalised: for the graph $y = -\frac{a}{x}$, the points $(1, -a)$ and $(-1, a)$ lie on the curve, because $-\frac{a}{(1)} = -a$ and $-\frac{a}{(-1)} = a$.

The graph of $y = -\frac{4}{x^2}$ has these properties:
- It has asymptotes at $x = 0$, $y = 0$.
- The points $(1, -4)$ and $(-1, -4)$ lie on the curve. This can be generalised: for the graph $y = -\frac{a}{x^2}$, the points $(1, -a)$ and $(-1, -a)$ lie on the curve, because $-\frac{a}{(1)^2} = -a$ and $-\frac{a}{(-1)^2} = -a$.

EXAMPLE 12

a) Complete the table of values for $y = \frac{2}{x}$ for $-4 \leq x \leq 4$. Round each value to 2 decimal places.

x	−4	−3	−2	−1	−0.5	0	0.5	1	2	3	4
$y = \frac{2}{x}$		−0.67		−2		−	4	2			0.5

b) Plot the graph of $y = \frac{2}{x}$ for $- \leq x \leq 5$.

c) Sketch the graph of $y = \frac{2}{x}$.

a)

x	−4	−3	−2	−1	−0.5	0	0.5	1	2	3	4
$y = \frac{2}{x}$	−0.5	−0.67	−1	−2	−4	−	4	2	1	0.67	0.5

b)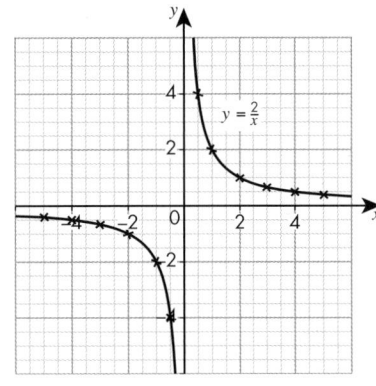

c) You know:
   - the shape of a reciprocal function
   - there are asymptotes at $x = 0$ and $y = 0$.

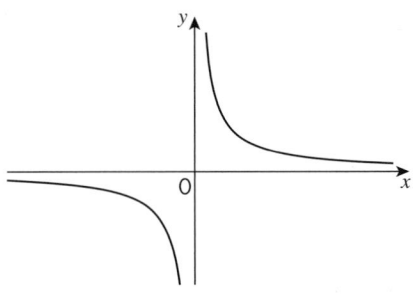

Drawing accurate graphs

These are some of the common errors in drawing accurate curves.

- When the points are too far apart, a curve tends to 'wobble'.
- Drawing curves in small sections leads to 'feathering'.
- The place where a curve should turn smoothly is drawn 'flat'.
- A line is drawn through a point that, clearly, has been incorrectly plotted.

Here are some tips that will make it easier for you to draw smooth, curved lines.

- If you are *right-handed*, turn your paper or exercise book round so that you draw from left to right. Your hand is steadier this way than when you are trying to draw from right to left or away from your body. If you are *left-handed*, you should find drawing from right to left the more accurate way.
- Move your pencil over the points as a practice run without drawing the curve.
- Do one continuous curve and only stop at a plotted point.
- Use a *sharp* pencil and do not press too heavily, so that you may easily rub out mistakes.

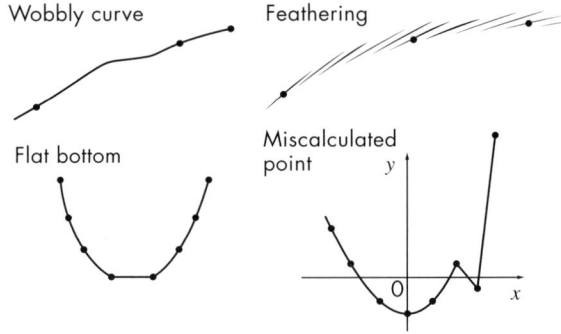

Transformations of exponential and reciprocal graphs

You can apply all the same transformations from **Sections 4.1** and **4.2** to exponential and reciprocal functions.

EXAMPLE 13

Apply the transformations to the graphs of $y = f(x)$ where:

i) $f(x + 2)$ ii) $-2f(x) - 3$ iii) $f(-3x)$

a) $f(x) = 3^x$ b) $f(x) = \frac{1}{x}$

a)
i) $f(x+2) = 3^{x+2}$

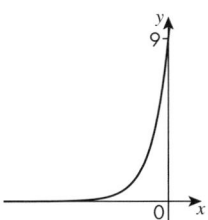

The graph is translated by the vector $\begin{pmatrix} -2 \\ 0 \end{pmatrix}$.
When $x = 0$, $3^{(0)+2} = 9$, so you can label the new y-intercept on the sketch.

ii) $-2f(x) - 3 = -2 \times 3^x - 3$

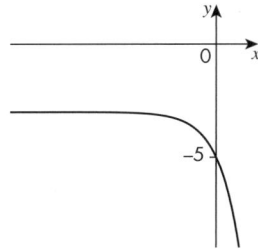

The graph is stretched in the y-direction by a factor of 2 and reflected in the x-axis. The -3 'outside' the function translates the graph by the vector $\begin{pmatrix} 0 \\ -3 \end{pmatrix}$.
$-2 \times 3^{(0)} - 3 = -2 \times 1 - 3 = -5$
Therefore, the new y-intercept is $(0, -5)$.

iii) $f(-3x) = 3^{-3x}$

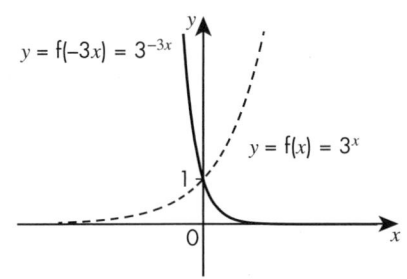

b)
i) $f(x+2) = \dfrac{1}{x+2}$

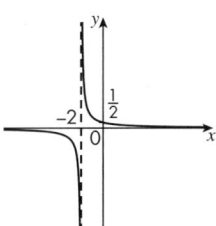

The graph is translated by the vector $\begin{pmatrix} -2 \\ 0 \end{pmatrix}$.
When $x = 0$, $\dfrac{1}{(0)+2} = \dfrac{1}{2}$, so you can label the new y-intercept on the sketch. The $f(x)$ asymptote $x = 0$ is translated to $x = -2$.

ii) $-2f(x) - 3 = -\dfrac{2}{x} - 3$

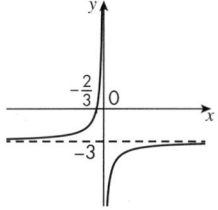

The graph is stretched in the y-direction by a factor of 2 and reflected in the x-axis. The -3 'outside' the function translates the graph by the vector $\begin{pmatrix} 0 \\ -3 \end{pmatrix}$.
The $f(x)$ asymptote $y = 0$ is translated to $y = -3$.
To find the x-intercept, set the transformed function equal to zero:
$-2f(x) - 3 = -\dfrac{2}{x} - 3 = 0$
$-\dfrac{2}{x} = 3$
$x = -\dfrac{2}{3}$

iii) $f(-3x) = \dfrac{1}{-3x}$

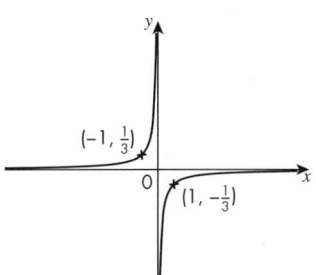

The graph is stretched in the *x*-direction by a factor of $\frac{1}{3}$ and reflected in the *y*-axis. The stretch can be difficult to see on a sketch, so $y = f(x) = 3^x$ has been included for reference.

The difference is clearer when considering individual points that are unnecessary for a sketch, e.g.
$x = 1$, $f(x) = 3^1 = 3$
$x = -1$, $f(-3x) = 3^{-3(-1)} = 27$
These points will be equidistant from the *y*-axis but have vastly different *y*-coordinates.

The graph is stretched in the *x*-direction by a factor of $\frac{1}{3}$ and reflected in the *y*-axis.

The point $(1, 1)$ has become $\left(-1, \frac{1}{3}\right)$ and the point $(-1, -1)$ has become $\left(1, -\frac{1}{3}\right)$.

EXAMPLE 14

The graph $y = ab^x$ passes through points A(0, 2) and B(1, 8).

a) Work out the value of *a* and the value of *b*.
b) Using your values of *a* and *b*, sketch the graph of $y = ab^x$.

a) $2 = ab^{(0)} = a \times 1 \Rightarrow a = 2$ $8 = 2b^{(1)} \Rightarrow 8 = 2b \Rightarrow b = 4$
b) $y = 2 \times 4^x$

You know:
- the shape of an exponential function
- that it crosses the *y*-axis at $(0, 2)$ because $2 \times 4^{(0)} = 2 \times 1 = 2$
- and that it has an asymptote at $y = 0$.

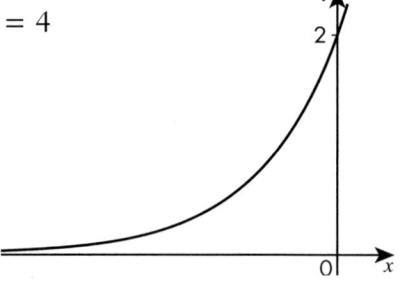

EXAMPLE 15

The graph $y = ab^x$ passes through points A(2, 45) and B(5, 1215). Work out the value of *a* and the value of *b*.

Use both coordinates to form equations in terms of *a*.

$45 = ab^{(2)}$ $1215 = ab^{(5)}$

$a = \frac{45}{b^2}$ $a = \frac{1215}{b^5}$

Equate the two equations and solve to find the value of *b*.

$\frac{45}{b^2} = \frac{1215}{b^5}$

$45b^5 = 1215b^2$

$\frac{b^5}{b^2} = \frac{1215}{45}$

$b^3 = 27$ so $b = 3$

HINTS AND TIPS

It doesn't matter which equation is used, but the other can be used to check that all working is correct.

Substitute b back into one of your starting equations to find the value of a.

$a = \dfrac{45}{(3)^2} = \dfrac{45}{9} = 5$

Check:

$\dfrac{1215}{(3)^5} = 5$ ✓

$a = 5$, $b = 3$

EXERCISE 4D

TARGETING PASS

1. **a)** Copy and complete the table of values for $y = \left(\dfrac{1}{3}\right)^x$ for $-2 \leq x \leq 3$. (Round each value to 2 dp.)

x	-2	-1	0	1	2	3
y	9		1	0.33		

   b) Plot the graph of $y = \left(\dfrac{1}{3}\right)^x$ for $-2 \leq x \leq 3$. (Take the y-axis from 0 to 10.)

   c) Use your graph to find an approximate solution to the equation $\left(\dfrac{1}{3}\right)^x = 2$.

2. **a)** Copy and complete this table of values for $y = \dfrac{5}{x}$. (Round each value to 2 dp.)

x	2	3	4	5	6
y			1.25		0.83

   b) Plot the graph of $y = \dfrac{5}{x}$ for $2 \leq x \leq 6$.

TARGETING DISTINCTION

3. The graph of $y = ab^x$ passes through points A(0, 3) and B(3, 81). Work out the value of a and the value of b.

4. The graph of $y = ab^x$ passes through points A(0, −2), B(2, −32) and C(−2, q). Work out the value q.

5. The graph of $y = ab^x$ passes through points A(2, 20) and B(5, 160). Work out the value of a and the value of b.

HINTS AND TIPS

Set up two simultaneous equations to solve, as in **Example 15**.

TARGETING DISTINCTION*

6. Sketch the graphs of $y = 5^x$ and $y = \left(\dfrac{1}{5}\right)^x$ on the same axes. Label the intercept of each graph.

7. Sketch the graphs of $y = \dfrac{1}{x}$, $y = \dfrac{2}{x}$ and $y = -\dfrac{2}{x}$ on the same axes.

8. Sketch the graphs of $y = \frac{1}{x^2}$ and $y = \frac{3}{x^2}$ on the same axes.

9. A sketch of the exponential graph of $y = f(x)$ is shown.

   The graph has an asymptote at $y = 1$ and a y-intercept of 2.

   Apply the following transformations, sketching each new graph on new axes.

   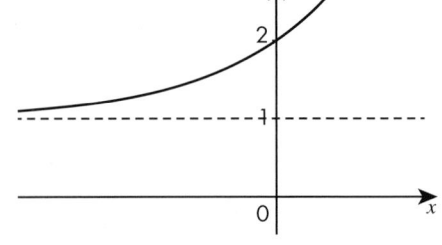

   a) $f(x+3)$ b) $f(2x)$ c) $-f(x)$ d) $f(-2x)+1$

10. A sketch of the reciprocal graph of $y = f(x)$ is shown. The graph has asymptotes at $x = 0$ and $y = 3$.

    Apply the following transformations, sketching each new graph on new axes. Label any asymptotes.

    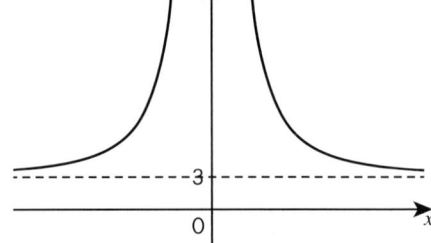

    a) $f(x) - 3$ b) $2f(x)$
    c) $f(-x)$ d) $-3f(x) - 1$

11. The graph $y = f(x)$, where $f(x) = 2^x$, is transformed so that it has an asymptote at $y = -1$ and an intercept of -3. Write down the equation of the transformed function.

4.5 Non-linear graphs

THIS SECTION WILL SHOW YOU HOW TO ...
✓ Interpret real-life graphs
✓ Calculate and estimate instantaneous and average rates of change
✓ Find and interpret the area under a graph
✓ Calculate the area under a graph using the trapezium rule.

KEY WORDS
✓ constant velocity   ✓ initial value   ✓ strip
✓ trapezium rule   ✓ distance–time graph   ✓ velocity–time graph

Interpreting real-life graphs

EXAMPLE 16

The graph shows the value of a car as it depreciates over time.

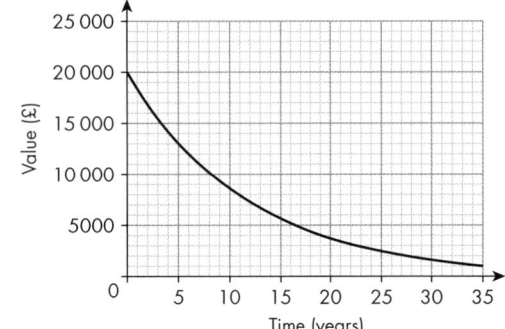

a) What is the **initial value** of the car?

b) What is the value of the car after 5 years?

c) After how many complete years has the car halved in value?

d) The graph suggests that after 30 years the car will be worth approximately £1500. Why might this be inaccurate?

a) When zero years have passed, the car is worth £20 000. This is the y-intercept of the graph.

b) ≈ £13 000

c) 20 000 ÷ 2 = £10 000

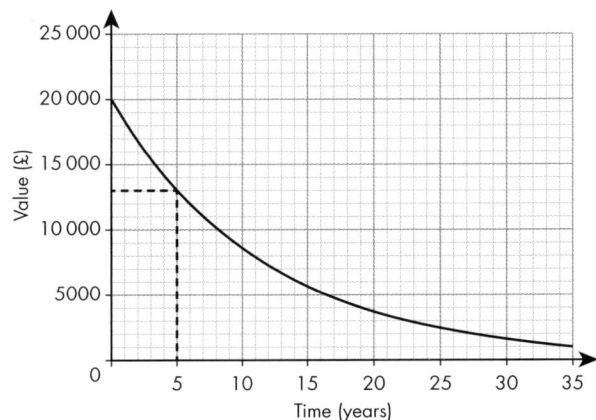

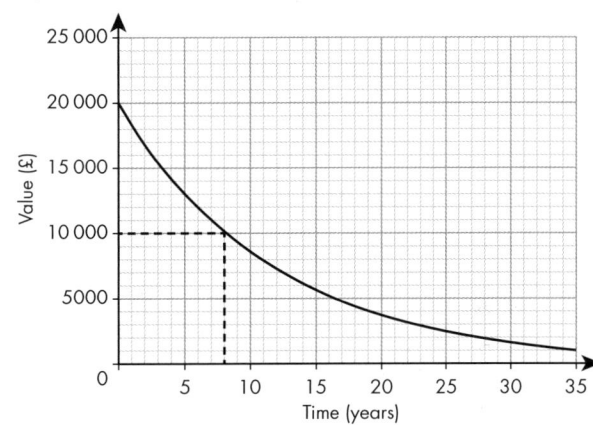

≈ 8 complete years.

d) A car is unlikely to still be drivable after 30 years.

Rates of change

You may also be interested in the rate at which the car's value is depreciating.

There are two ways of measuring this:
- Instantaneous rate of change
- Average rate of change.

To calculate the instantaneous rate of change at a given point, you need to calculate an estimate for the gradient at that point. When a graph is curved, you need to draw a tangent to the graph at the point of interest and then find the gradient of the tangent.

To calculate the average rate of change, you draw a straight line connecting the two points of interest and find the gradient of this line.

EXAMPLE 17

Look again at the graph in **Example 16**.

a) Estimate the rate of depreciation of the car after 5 years. Interpret your answer.

b) Estimate the rate of depreciation after 10 years.

c) Estimate the average rate of depreciation between 5 and 10 years.

a) Draw a tangent to the curve at the point $(5, \approx 13000)$.

Find the gradient of the tangent using the method from **Chapter 3.1**:

$$m = \frac{\Delta y}{\Delta x} = \frac{8000}{7.5} = 1066.67$$

After 5 years, the car is depreciating at a rate of £1066.67 per year.

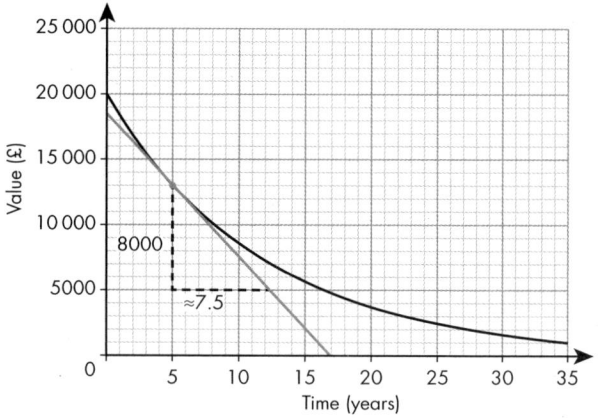

b) Draw a tangent to the curve at the point $(10, \approx 8700)$.

Find the gradient of the tangent:

$$m = \frac{\Delta y}{\Delta x} = \frac{8000}{11} = 727.273$$

After 10 years, the car is depreciating at a rate of £727.27 per year.

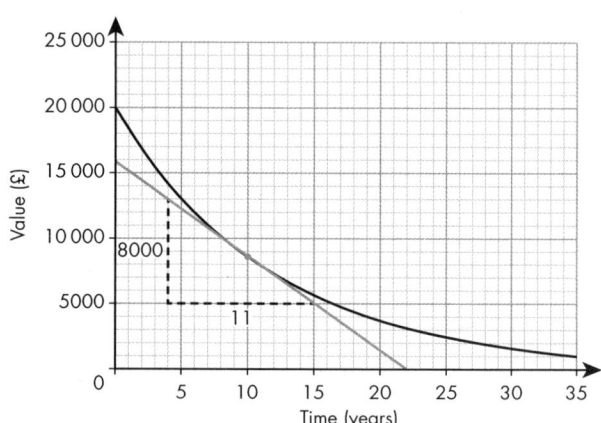

c) Draw a line segment connecting the points $(5, \approx 13000)$ and $(10, \approx 8700)$.

Find the gradient of the line segment:

$$m = \frac{\Delta y}{\Delta x} = \frac{4300}{5} = 860$$

Between 5 and 10 years, the car depreciates at an average rate of £860 per year.

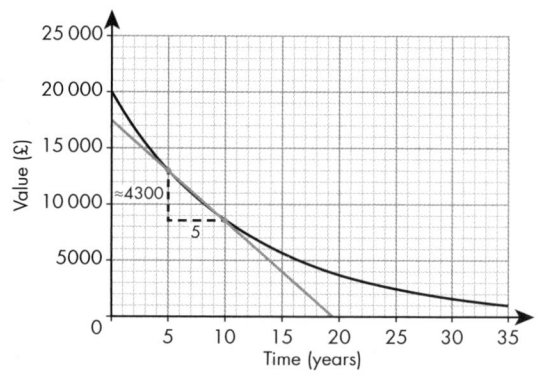

Average rates of change can also be calculated from straight-line graphs.

EXAMPLE 18

The graph shows the height of a toy plane over a 50-second period.

a) After how many seconds is the plane 20 m above the ground?

b) Calculate the rate at which the height of the plane is increasing over the first 10 seconds.

c) Describe what is happening to the plane from 10 to 50 seconds.

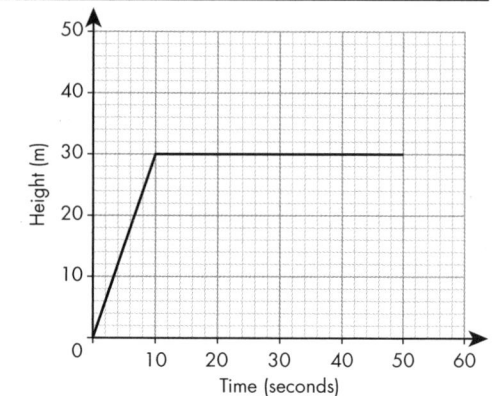

a)

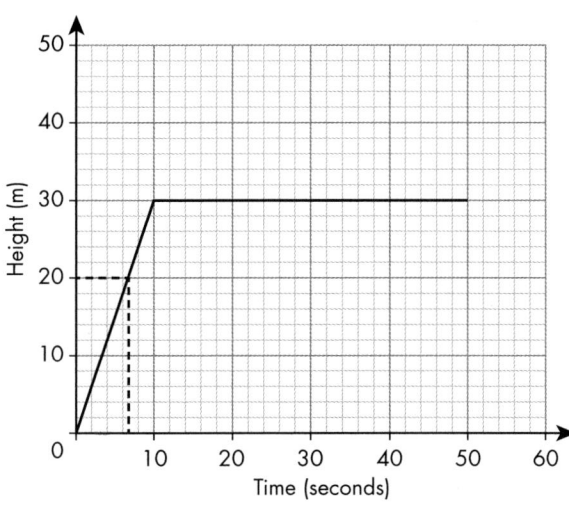

Approximately 7 seconds.

b)
Find the gradient of the line segment from (0, 0) to (10, 30).

$$m = \frac{\Delta y}{\Delta x} = \frac{30}{10} = 3 \text{ m/s}$$

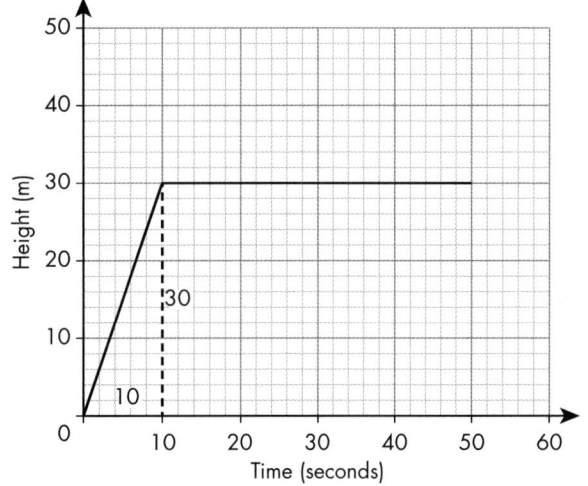

c) The toy plane is flying at a constant height of 30 m.

Two common graphs that you will encounter are **distance–time graphs** and **velocity–time graphs**.

On a distance–time graph, the gradient of the line or curve gives the speed at which the object is travelling.

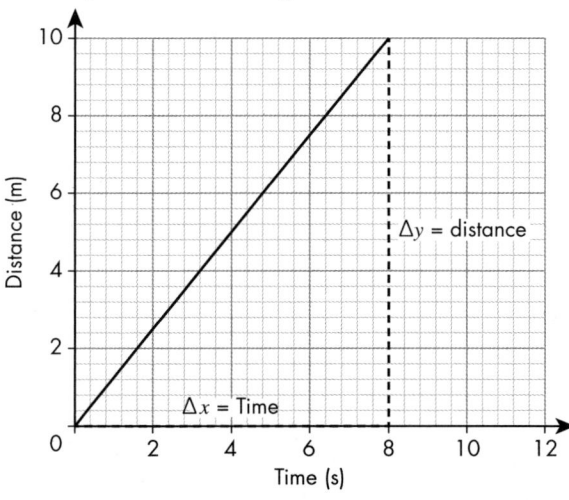

Finding the gradient is related to the formula for calculating speed:

$$m = \frac{\Delta y}{\Delta x} = \frac{D}{T} = S$$

On a velocity–time graph, the gradient of the line or curve gives the acceleration or deceleration of the object. Acceleration is the rate of change of velocity.

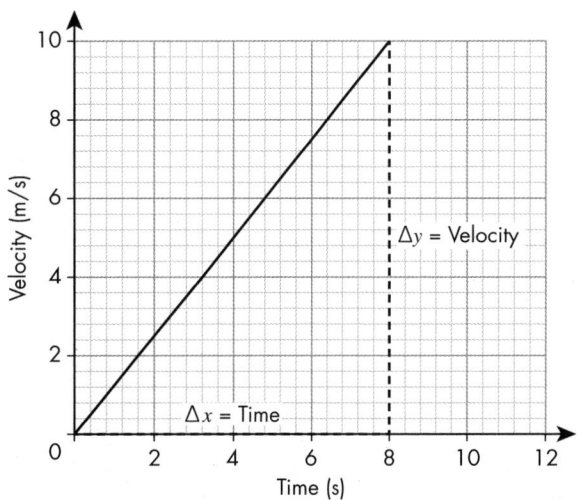

Finding the gradient of a velocity–time graph enables you to deduce the units of acceleration, m/s^2.

$$m = \frac{\Delta y}{\Delta x} = \frac{m/s}{s} = m/s/s = m/s^2$$

Area under curves

For a velocity–time graph, the distance travelled is calculated by finding the area under the curve.

In this graph, an object is moving at a **constant velocity** of 5 m/s.

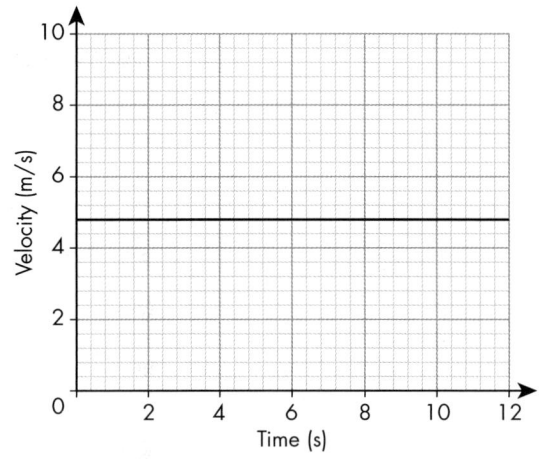

The formula connecting speed, distance and time is $S = \frac{D}{T}$. Therefore, $D = S \times T$. To find the distance travelled in the first 10 seconds, you can draw a vertical line from the x-axis to the point (10, 5) to create a rectangle with a height representing 5 m/s and a base representing 10 s. The distance travelled is the area of this rectangle:

Area $= B \times H = 10 \times 5 = S \times T = D$

Therefore, the distance travelled is $D = 10 \times 5 = 50$ m

Adding units to the calculation shows why the answer is in metres (not 'units²'):

$5\frac{m}{s} \times 10 \text{ s} = 5\frac{m}{s} \times 10 \text{ s} = \frac{50 \text{ ms}}{s}$ (the 's' cancel out)
$= 50$ m

EXAMPLE 19

The graph shows the velocity of a car as it accelerates.

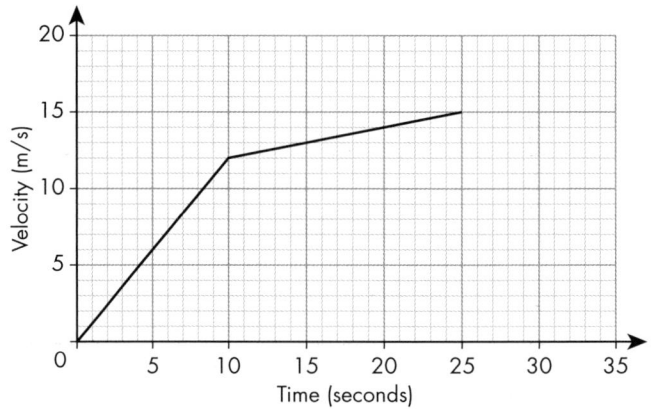

Calculate the distance travelled by the car in the first 25 seconds.

Split the area under the graph into a triangle and a trapezium:

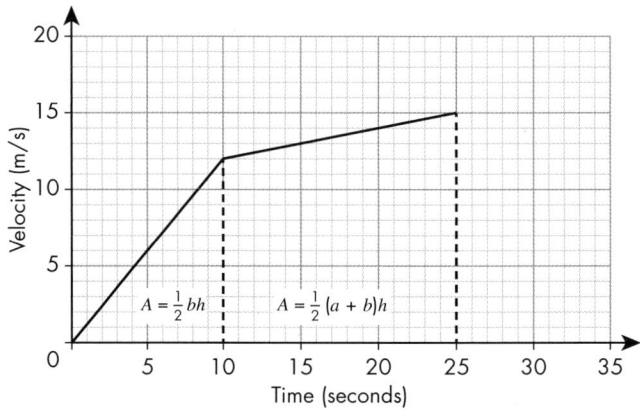

Triangle
$A = \frac{1}{2}bh = \frac{1}{2} \times 10 \times 12 = 60$

Trapezium
$A = \frac{1}{2}(a+b)h = \frac{1}{2}(12+15) \times 15 = 202.5$

Distance = 60 + 202.5 = 262.5 m

HINTS AND TIPS

The area under the graph can be split in any way that allows you to calculate the area. For example:

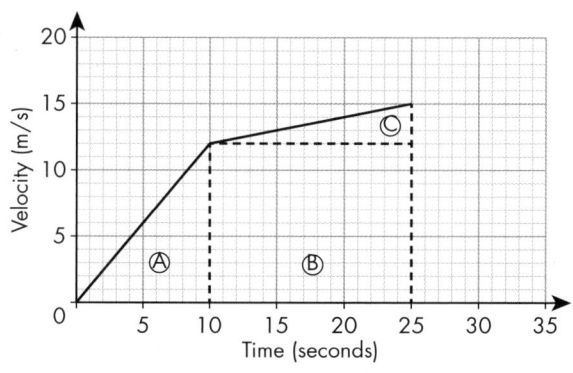

Finding the area under a graph becomes more challenging when the graph is curved.

The diagram shows part of the graph of
$y = (x+3)(x-2)(x+1)$:

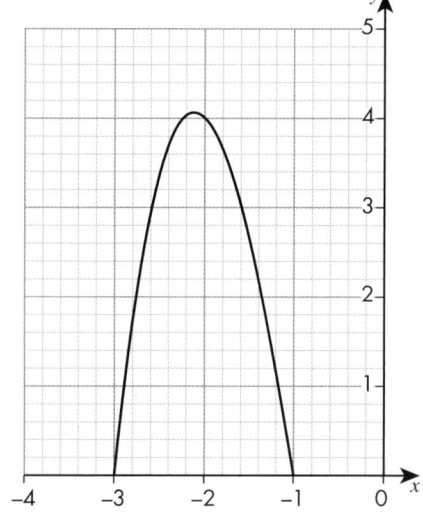

You could estimate the area under this curve in several ways.

Two triangles:

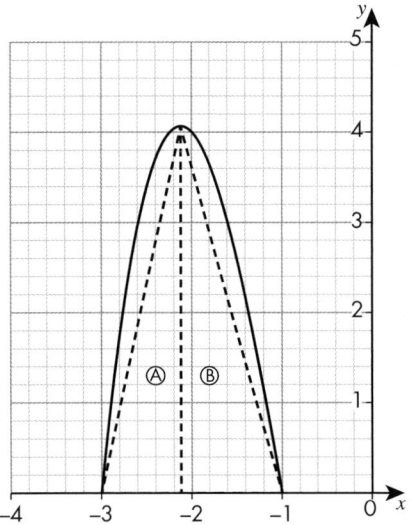

However, this leaves out a large part of the area, so the calculation would underestimate the true area considerably.

Two triangles and two trapeziums:

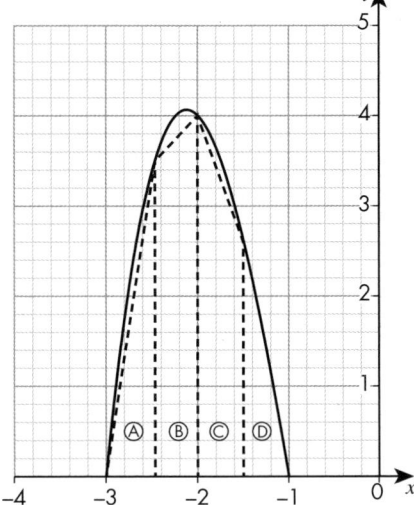

There are still parts of the area unaccounted for, although the answer will be closer to the actual area than using just two triangles.

HINTS AND TIPS

When dividing the area under a curved graph, it is usually best to use shapes (often, in this context called '**strips**') of equal width. Although this is not strictly necessary at this stage, it is essential for the **trapezium rule** in the next section.

Two triangles and three trapeziums:

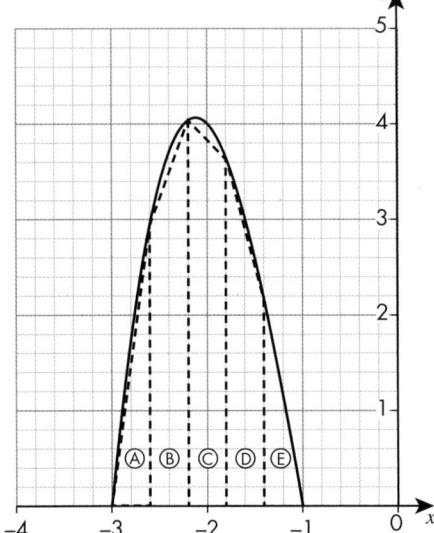

As the strips get smaller, the estimate will become more accurate as the area covered gets closer to the actual area under the curve.

Two triangles and eight trapeziums:

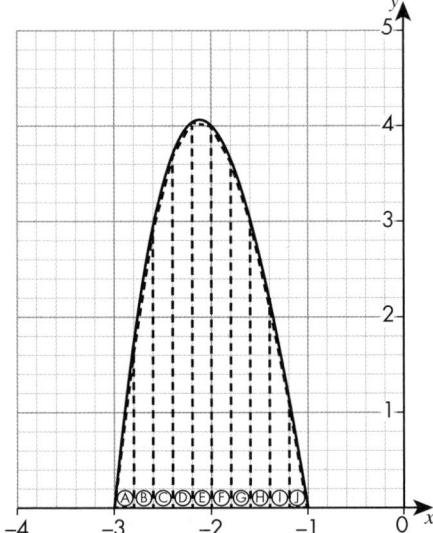

However, although it is more accurate, using a larger number of strips adds more terms to the calculation of the area.

You will usually be told how many strips to use in the question.

EXAMPLE 20

The diagram shows part of the graph of $y = \frac{1}{x}$.

a) Find the area under the graph between $x = 0.5$ and $x = 2.5$. Use four strips of equal width. Give your answer to two decimal places.

b) State whether your answer in part **a** is an overestimate or an underestimate.

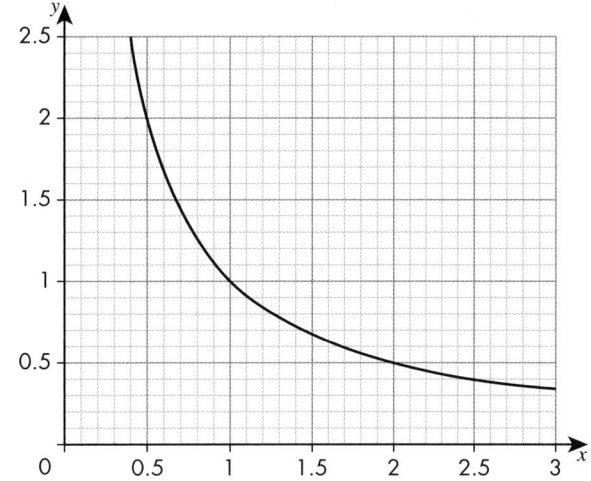

a) Splitting the area into four strips gives four trapeziums

(Area = $\frac{1}{2}(a + b)h$)

A: $\frac{1}{2}(2 + 1) \times 0.5 = 0.75$

B: $\frac{1}{2}(1 + 0.67) \times 0.5 = 0.4175$

C: $\frac{1}{2}(0.67 + 0.5) \times 0.5 = 0.2925$

D: $\frac{1}{2}(0.5 + 0.4) \times 0.5 = 0.225$

Area = $0.75 + 0.4175 + 0.2925 + 0.225 = 1.685 = 1.69$ units² (to 2dp)

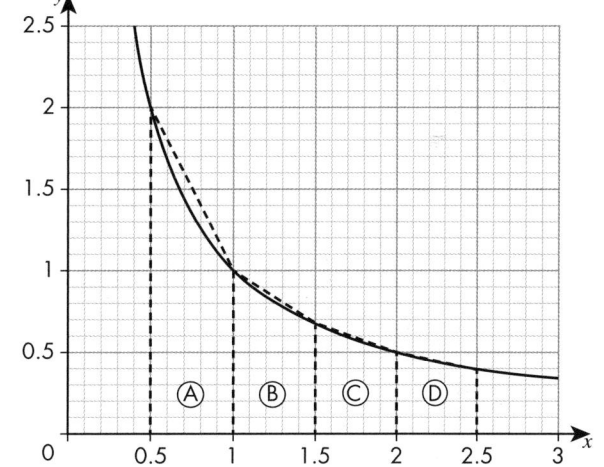

b) 1.69 is an overestimate as the trapeziums go above the curve.

The trapezium rule

To estimate the area under a curve more efficiently, you can use the trapezium rule.

The formula for the trapezium rule is:

Area = $\frac{1}{2}h(y_0 + 2(y_1 + y_2 + \ldots + y_{n-1}) + y_n)$,

where h is the width of each strip.

Here is the graph from **Example 20** again with the area divided into four strips of equal width.

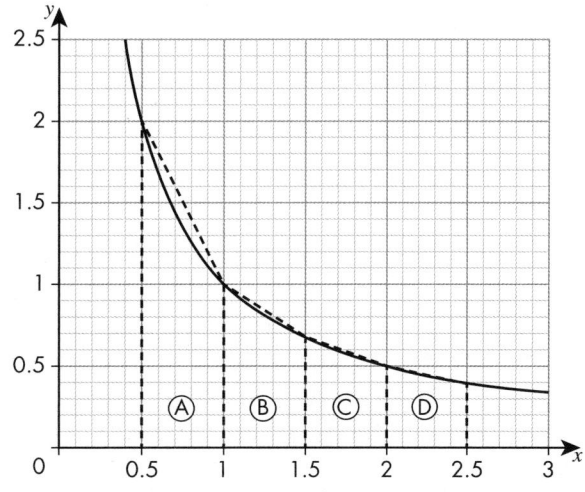

Consider trapezium A.

The trapezium has width 0.5. Therefore, $h = 0.5$.

When using the trapezium rule, the parallel sides of trapezium A (a and b in the formula for the area of a trapezium), are labelled y_0 and y_1. This labelling continues for each trapezium:

The total area is the area of each trapezium added together, as you saw in **Example 20**.

Using the notation in the diagram above, this becomes:

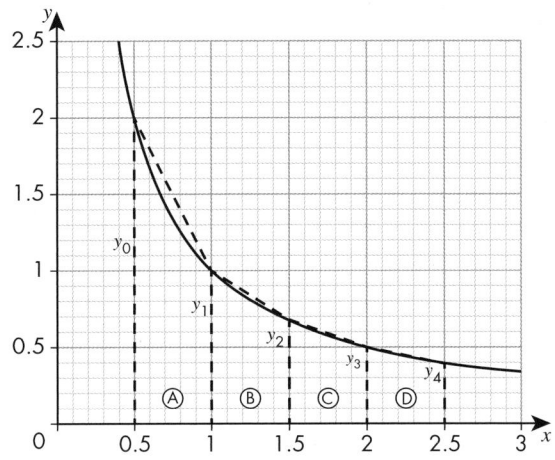

$$\text{Area} = \tfrac{1}{2}(y_0 + y_1)h + \tfrac{1}{2}(y_1 + y_2)h + \tfrac{1}{2}(y_2 + y_3)h + \tfrac{1}{2}(y_3 + y_4)h$$

$$= \tfrac{1}{2}h(y_0 + y_1 + y_1 + y_2 + y_2 + y_3 + y_3 + y_4)$$

$$= \tfrac{1}{2}h(y_0 + 2(y_1 + y_2 + y_3) + y_4)$$

Substituting the heights and width of the trapeziums into this formula gives the total area:
$\text{Area} = \tfrac{1}{2} \times 0.5 \times (2 + 2(1 + 0.67 + 0.5) + 0.4) = 1.685 = 1.69$ units^2 (2 dp.).

In the formula for the trapezium rule (Area = $\tfrac{1}{2}h(y_0 + 2(y_1 + y_2 + \ldots + y_{n-1}) + y_n)$),

the heights y_n and y_{n-1} denote the final y-value and the penultimate y-value. In the example above, these are y_3 and y_4.

The estimate found using the trapezium rule becomes more accurate when the area is divided into a greater number of thinner strips (as $h \to 0$). You will usually be told how many strips to use in the question.

EXAMPLE 21

The diagram shows part of the graph of $y = \tfrac{1}{3} \times 2^x + 1$.

Use the trapezium rule with:

a) 4 strips

b) 8 strips

to estimate the area under the graph between $x = 0$ and $x = 4$.

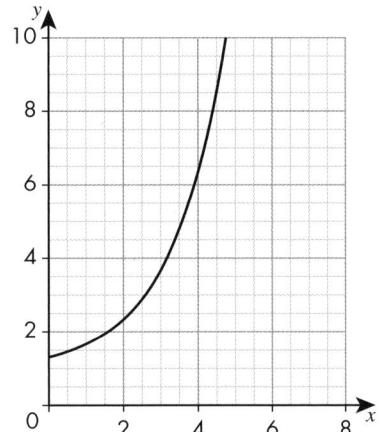

a) Using 4 strips

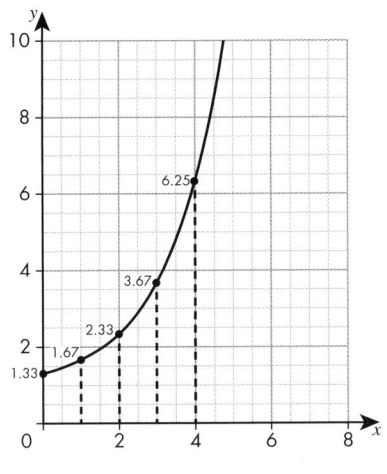

Trapezium rule

Area = $\frac{1}{2}h(y_0 + 2(y_1 + y_2 + \ldots + y_{n-1}) + y_n)$

$h = 1$. $y_0, y_1 \ldots y_4$ are marked on the diagram.

Area = $\frac{1}{2} \times 1 \times (1.33 + 2(1.67 + 2.33 + 3.67) + 6.25)$
     = 11.46

b) Using 8 strips

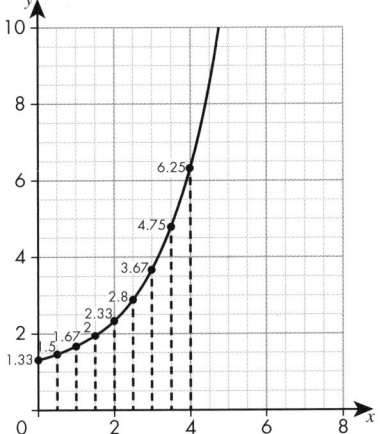

Trapezium rule

Area = $\frac{1}{2}h(y_0 + 2(y_1 + y_2 + \ldots + y_{n-1}) + y_n)$

$h = 0.5$. $y_0, y_1 \ldots y_9$ are marked on the diagram.

Area = $\frac{1}{2} \times 0.5 \times (1.33 + 2(1.5 + 1.67 + 2 + 2.33 + 2.8 + 3.67 + 4.75) + 6.25)$
     = 11.255

You will not always be given the graph on which to draw the values for y_0, y_1 etc.

To calculate the value of h, you can use formula: $h = \frac{y_0 + y_n}{n}$, where n represents the number of strips used.

Example 22 looks at the same question again without the diagram.

EXAMPLE 22

Use the trapezium rule with:

a) 4 strips

b) 8 strips

to estimate the area under the graph of $y = \frac{1}{3} \times 2^x + 1$ between $x = 0$ and $x = 4$.

a)

$y_0 = 0$, $y_4 = 4$

$\frac{0+4}{4} = 1$

$h = 1$

Create a table of values and substitute the x-values into $\frac{1}{3} \times 2^x + 1$.

x	0	1	2	3	4
y_n	y_0	y_1	y_2	y_3	y_4
$\frac{1}{3} \times 2^x + 1$	1.33	1.67	2.33	3.67	6.33

Then substitute these values into the formula for the trapezium rule:

Area = $\frac{1}{2} \times 1 \times (1.33 + 2(1.67 + 2.33 + 3.67) + 6.33) = 11.5$

b)

$y_0 = 0$, $y_8 = 4$

$\frac{0+4}{8} = 0.5$

$h = 0.5$

x	0	0.5	1	1.5	2	2.5	3	3.5	4
y_n	y_0	y_1	y_2	y_3	y_4	y_5	y_6	y_7	y_8
$\frac{1}{3} \times 2^x + 1$	1.33	1.47	1.67	1.94	2.33	2.89	3.67	4.77	6.33

Area = $\frac{1}{2} \times 0.5 \times (1.33 + 2(1.47 + 1.67 + 1.94 + 2.33 + 2.89 +$

$3.67 + 4.77) + 6.33) = 11.285$

EXERCISE 4E

TARGETING PASS

1. The speed–time graph shows the journey of a toy car over a 10-second period.

   a) Describe how the toy car is travelling between 2 and 6 seconds.

   b) What was the toy cars peak speed? Give your answer in km/h.

   c) Calculate the acceleration of the toy car over the first 2 seconds.

   > **HINTS AND TIPS**
   >
   > Remember that the units for acceleration are m/s² or ms⁻².

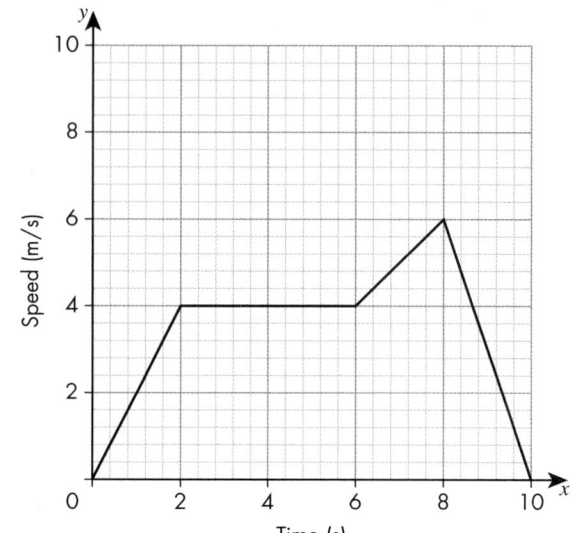

d) Calculate the distance covered by the toy car across the 10-second period.

e) Without further calculation, state how you know that the car decelerates in the final 2 seconds at a faster rate than it accelerated.

TARGETING MERIT

2. The graph shows the number of people affected by a disease for 70 days after it is first detected.

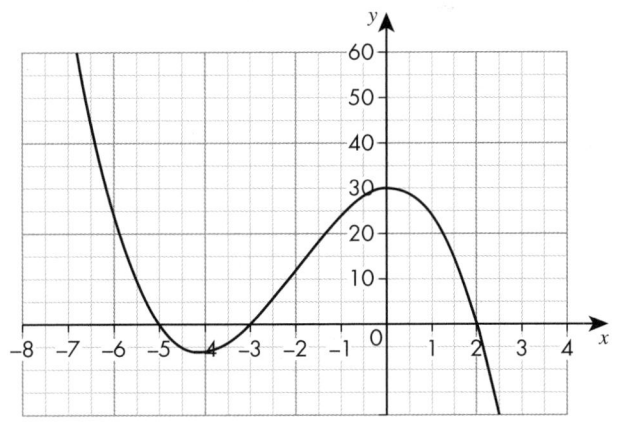

a) How many cases were initially detected?

b) The graph has an asymptote. What does this tell you about the disease?

c) Estimate the rate of infection:

   i) after 20 days

   ii) after 40 days.

d) Comment on what your answers to **c)** tell you about the disease.

e) Estimate the average rate of infection across the first 70 days.

TARGETING DISTINCTION

3. a) Sketch the curve of $y = x(4-x)$ for value of x from 0 to 4.

b) Estimate the gradient of the curve at $x = 3$.

c) Using 4 strips, estimate the area under the curve between $x = 0$ and $x = 4$.

TARGETING DISTINCTION*

4. The diagram shows the graph of $y = f(x)$.

a) Estimate the gradient of the curve at $x = -2$.

b) Using the trapezium rule with 5 strips, estimate the area under the curve between $x = -3$ and $x = 2$.

c) The graph is transformed so that $y = f(-x)$. State an estimate for the area under the curve between $x = -2$ and $x = 3$.

5. Use the trapezium rule with:

a) 3 strips

b) 6 strips.

to estimate the area under the graph of $y = \frac{1}{10}(x+6)^3(2-x)$ between $x = -5$ and $x = 1$.

6. Use the trapezium rule with:

   a) 4 strips

   b) 8 strips.

   to estimate the area under the graph of $y = x - \frac{2}{x}$ between $x = 3$ and $x = 7$. Give your answers to 1 decimal place.

7. a) Sketch the graph of $y = 2^x + 1$, marking the y-intercept and the equation of the horizontal asymptote.

   b) Use the trapezium rule with 4 strips to find an estimate for the area under the curve between $x = -3$ and $x = 1$, giving your answer to 1 decimal place.

Exam-style questions

1. The graph $y = f(x)$ passes through the points $A(3, -2)$ and $B\left(-\frac{1}{3}, 7\right)$.

   State the coordinates of points A and B after the following transformations.
   a) $y = 2f(x + 2)$ [2 marks]
   b) $y = f(-x) - 3$ [2 marks]

2. a) Sketch the graph of $y = f(x)$, where $f(x) = 2x^2 - x - 6$. [3 marks]
   b) One separate axes, sketch the graph of $y = f(-2x)$. [2 marks]

3. A circle has the equation $(x + 4)^2 + (y - 2)^2 = 10$.
   a) Show that the point $P(-1, 1)$ lies on the circle. [1 mark]
   b) Find the equation of the tangent to the circle at the point P. [4 marks]

4. The velocity–time graph shows the velocity of a car.

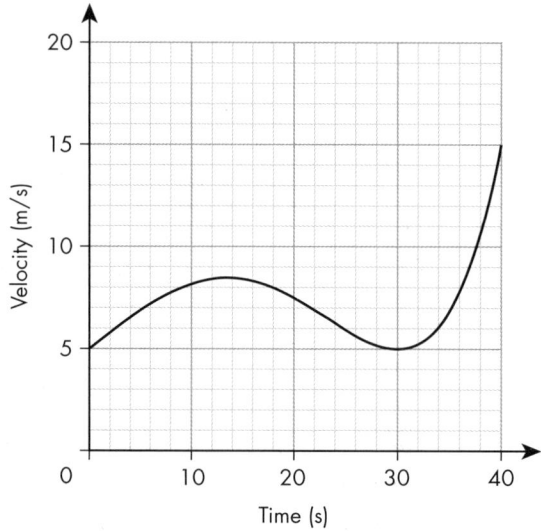

   a) State a time when the car was not accelerating or decelerating. [1 mark]
   b) Calculate the deceleration of the car at 20 seconds. [2 marks]

c) Calculate the average acceleration of the car in the first 40 seconds. [2 marks]

d) Using 4 strips, estimate the distance travelled by the car in the first 40 seconds. [3 marks]

5. A circle C has the equation $x^2 + y^2 - x + 6y = \frac{23}{4}$.

   a) Find the centre and radius of circle C. [4 marks]

   The points A and B are where circle C intersects the y-axis.

   b) Show that the length of AB can be written as $\sqrt{p}$, where p is an integer. [4 marks]

6. Sketch the graph of $y = \frac{3}{x+2}$. Label the asymptotes and the point where the curve crosses the y-axis. [4 marks]

7. The graph of $y = ab^x$ passes through points $A(-2, 18)$ and $B\left(3, \frac{2}{27}\right)$. Work out the value of a and the value of b. [4 marks]

8. The graph of $y = f(x)$, where $f(x) = \sin x$, has a maximum point at $Q(450°, 1)$.

   The transformation $y = af(x + b)$ moves point Q from $(450°, 1)$ to $(405°, 3)$.

   Work out the value of a and the value of b. [2 marks]

9. Use the trapezium rule with 4 strips to estimate the area of the region under the graph $y = \sqrt{3 + x}$ between $x = -2$, $x = 2$ and the x-axis.

   Give your answer to 3 decimal places. [4 marks]

5 Functions

5.1 Functions

THIS SECTION WILL SHOW YOU HOW TO …
- ✓ understand what functions are
- ✓ understand and use set and function notation, for example f(x), $\mathbb{R}$ and $\mathbb{Z}$
- ✓ understand how to identify the domain and range of a function from its equation and from its graph
- ✓ understand how to identify values that must be excluded from the domain of a function
- ✓ substitute numbers into a function, knowing that, for example f(2) is the value of the function when $x = 2$

KEY WORDS
- ✓ function
- ✓ mapping
- ✓ many-to-one
- ✓ domain
- ✓ roots
- ✓ element
- ✓ range
- ✓ one-to-one
- ✓ asymptote

A **function** is a relationship between two sets of values.

Equations written in terms of x and y such as $y = 3x - 4$ or $y = 2x^2 + 5x - 3$ show that y is a function of x. The value of y depends on the value of x.

Alternatively, you can use a different notation, writing f(x) = $3x - 4$ and calling it 'function f'. Then you can show the result of using different values for x, for example:

- 'the value of f(x) when x is 5' can be written as f(5).
  So f(5) = $3 \times 5 - 4 = 11$.
- f(1) means 'the value of f(x) when x is 1'.
  So f(1) = $3 \times 1 - 4 = -1$ and f(-1) = -7.
- If there are different functions in the same problem, you can use different letters, for example, g(x) = $2x^2 + 5x - 3$ or 'function g'.

Consider function f(x) = $\sqrt{x - 3}$.

f(3) = $\sqrt{3 - 3}$ = 0
f(4) = $\sqrt{4 - 3}$ = 1
f(19) = $\sqrt{19 - 3}$ = $\sqrt{16}$ = 4
f(103) = $\sqrt{103 - 3}$ = $\sqrt{100}$ = 10

This function shows a connection between two sets of numbers. The starting set is called the **domain**.

The resulting set of numbers is called the **range**.

This connection between the two sets is called a **mapping**.

You cannot find the square root of a negative number so numbers less than 3 must be excluded from the domain of f in this case.

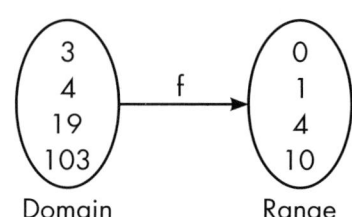

Domain — Range

EXAMPLE 1

$g(x) = \frac{1}{x}$

a) What number must be excluded from the domain of g?

b) If the domain is $\{x: 1 \leq x \leq 2\}$, what is the range of $y = g(x)$?

a) You cannot evaluate $\frac{1}{0}$, so $x = 0$ must be excluded from the domain of g.

b) $g(1) = \frac{1}{1} = 1 \qquad g(2) = \frac{1}{2}$

The range will be all the numbers from $\frac{1}{2}$ to 1. You can write this in set notation as $\{y: \frac{1}{2} \leq y \leq 1\}$.

EXAMPLE 2

$f(x) = x^2$ has domain $x > 5$.
State the range of $f(x)$.
$f(x) > 5^2$, so the range of $f(x)$ is $f(x) > 25$.

HINTS AND TIPS

The domain is the x values, or set of possible inputs, and the range is the possible outputs of a function.

EXAMPLE 3

$f(x) = 3x - 2$ and $f(x) > 4$.
Work out the domain of this function.
$3x - 2 > 4$
$\quad 3x > 6$
So, the domain is $x > 2$.

EXERCISE 5A

TARGETING PASS

1. $h(x) = \frac{12}{x} + 1 \qquad k(x) = 2^x - 1$
   a) Work out $h(6)$.　　　　　**b)** Work out $k(-1)$.

2. What values of x must be excluded from the domains of these functions?
   a) $f(x) = \sqrt{x}$　　**b)** $g: = \frac{1}{x+1}$　　**c)** $h: = \sqrt{x+1}$
   d) $j: = \frac{1}{2x+1}$　　**e)** $k(x) = \frac{1}{x^2 - 3x + 2}$

3. $f(x) = x^2 + 1$. Work out the range for each domain.
   a) $\{3, 4, 5\}$　　**b)** $\{-2, -1, 0, 1, 2\}$　　**c)** $\{x: 1 \leq x \leq 2\}$
   d) $\{x: x \geq 10\}$　　**e)** $\{x: x \leq -10\}$

4. Each of the following functions has domain {1, 2, 3, 4}. Work out the range for each function.

   a) $f(x) = (x - 2)^2$ b) $g(x) = \frac{1}{x}$ c) $h(x) = 2x + 3$

   d) $f(x) = 6 - x$ e) $g(x) = (x - 1)(x - 4)$

5. $f(x) = x^2$. Explain why −2 could be in the domain but cannot be in the range.

6. The domain of a function f is {1, 2, 3, 4} and the range is {2, 3, 4, 5}. State whether or not each of these is a possible description of f:

   a) $f(x) = x + 1$ b) $f(x) = 2x$ c) $f(x) = 6 - x$

7. $f(x) = 2x - 9$ and $f(x) > 0$. State the smallest whole number value of x.

8. $f(x) = 3x + 1$ and $x > 5$

   a) State the range of $f(x)$.   b) Work out the domain and range of $f(4x)$.

9. $f(x) = x^2 + 3$ and $x > 0$

   a) State the range of $f(x)$.   b) Work out the domain and range of $f(2x)$.

TARGETING MERIT

10. $f(x)$ is a quadratic graph of the form $y = x^2 + c$.

    The range of $f(x)$ is $f(x) > 7$.

    a) Write down the value of c.   b) Work out $f(2)$.

11. The function $f(x)$ is defined as $f(x) = 11 - 2x$ for $a < x < b$.

    The range of $f(x)$ is $-5 < f(x) < 7$. Work out the values of a and b.

Function notation and graphs

$\mathbb{R}$ denotes the set of real numbers which includes all possible number values (including rational and irrational numbers).

$\mathbb{Z}$ denotes the set of all integers $\{\ldots -2, -1, 0, 1, 2, 3 \ldots\}$.

$\{x \in \mathbb{R}\}$ means that x is a member of the set of real numbers.

The **roots** of an equation $y = f(x)$ are the values of x for which $f(x) = 0$.

A **one-to-one** function maps one element in the domain to one element in the range.

A **many-to-one** function maps more than one **element** in the domain to one element in the range.

One-to-many and many-to-many mappings are not functions.

A function can be defined as a set of distinct parts, each with its own domain. For any function, one element from the domain maps to exactly one (and only one) element of the range.

EXAMPLE 4

A function f(x) is defined as
$f(x) = x^2 \quad\quad 0 \leqslant x < 2$
$\quad\quad = 4 \quad\quad\quad 2 \leqslant x < 3$
$\quad\quad = 7 - x \quad\; 3 \leqslant x \leqslant 5$

Draw the graph of y = f(x) on the grid.

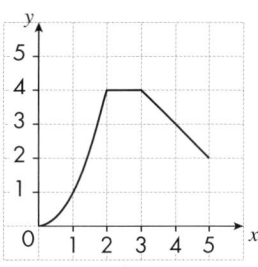

EXAMPLE 5

The graph of y = f(x) is shown.
Define f(x), stating the domain for each part.
$f(x) = 2x \quad\quad 0 \leqslant x < 2$
$\quad\quad = 6 - x \quad\; 2 \leqslant x < 4$
$\quad\quad = 2 \quad\quad\quad 4 \leqslant x \leqslant 5$

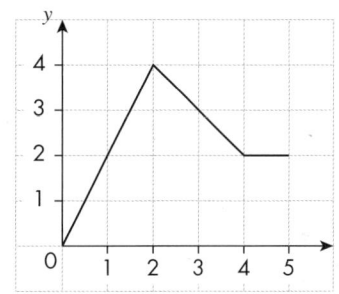

EXAMPLE 6

a) What kind of function is f(x) = 3x − 1.

a) This is a one-to-one function as every element in the domain maps to one element in the range.

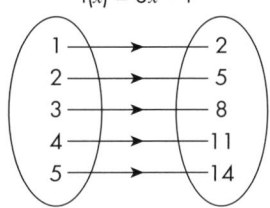

b) What kind of function is $f(x) = 2x^2$?

b) This is a many-to-one function as more than one element in the domain maps to one element in the range.

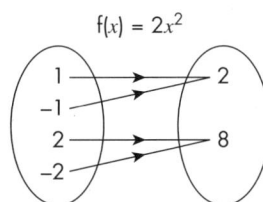

EXAMPLE 7

The graph of $y = f(x)$ is shown below with a domain $\{x \in \mathbb{R}: -4 \leq x \leq 2\}$.
Write down the range of f(x).

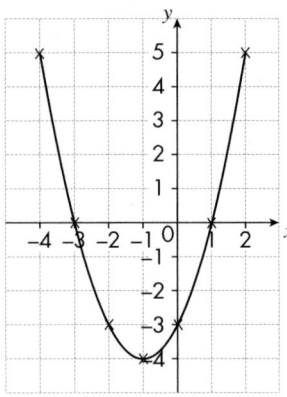

> **HINTS AND TIPS**
>
> This is an example of a many-to-one function as f(−4) = 5 *and* f(2) = 5.
> $\{x \in \mathbb{R}\}$ and $\{f(x) \in \mathbb{R}\}$ are the domain and range of a function where the domain or range has not been restricted.

The domain of the function f(x) is defined as the set of real numbers from −4 to 2.

The graph shows that the range of f(x) is $\{f(x) \in \mathbb{R}: -4 \leq f(x) \leq 5\}$, which is all the real values from −4 to 5.

EXERCISE 5B

TARGETING PASS

1. In each part, sketch the graph of $y = f(x)$.

   a) $f(x) = 2$ $0 \leq x < 2$
        $= x$ $2 \leq x < 4$
        $= 4$ $4 \leq x \leq 6$

   b) $f(x) = x$ $0 \leq x < 2$
        $= 2$ $2 \leq x < 4$
        $= 6 - x$ $4 \leq x \leq 6$

   c) $f(x) = x + 1$ $0 \leq x < 2$
        $= 3$ $2 \leq x < 4$
        $= x - 1$ $4 \leq x \leq 6$

   d) $f(x) = x^2$ $0 \leq x < 2$
        $= 8 - 2x$ $2 \leq x \leq 4$

   e) $f(x) = 2 + x$ $0 \leq x < 2$
        $= 4$ $2 \leq x < 4$
        $= 6 - \frac{1}{2}x$ $4 \leq x \leq 6$

   f) $f(x) = x^2$ $0 \leq x < 1$
        $= x$ $1 \leq x < 3$
        $= 2x - 3$ $3 \leq x \leq 6$

2. Write down the range of f(x) for each part of question 1.

3. Graphs of $y = f(x)$ are shown.

   Define $f(x)$ stating the domain for each part of each graph.

   a)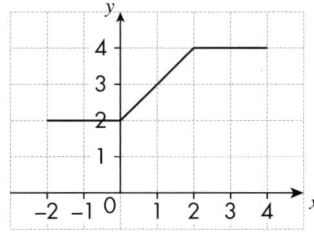

   b)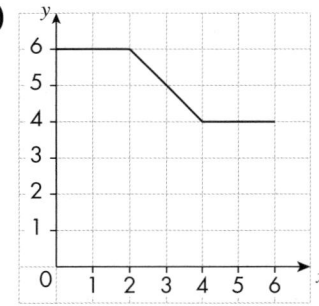

   c)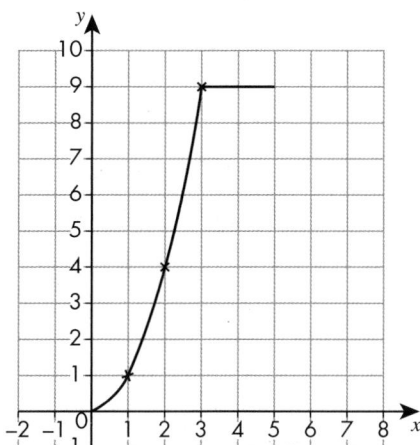

   d)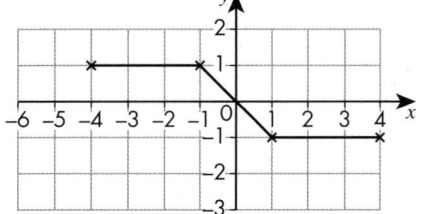

TARGETING DISTINCTION

4. Which of these graphs does not represent a function?
   Give a reason for your answer.

   a)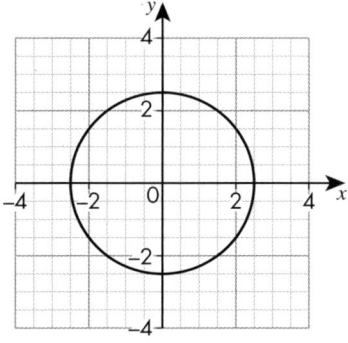

   b)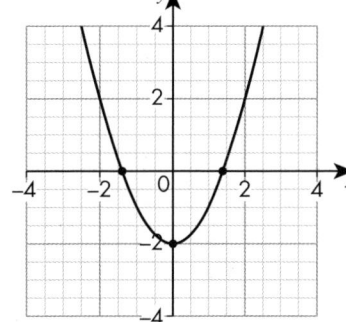

   c)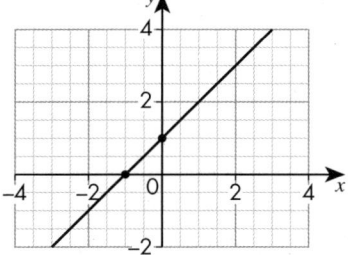

   d)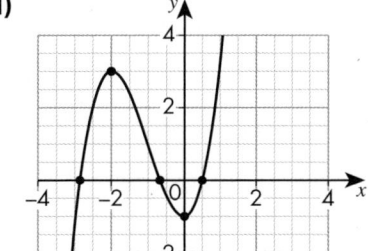

5. a) Sketch the graph of f(x) = 2x − 1 for values of x between −2 and 4.
   b) Define the domain.
   c) Define the range.

EXAMPLE 8

This is the graph of $f(x) = \dfrac{2}{1-x}$.

Define the domain and the range of f(x).

The line x = 1 is an **asymptote** and so x cannot equal 1. Also, f(x) cannot take the value zero as there are no values of x for which f(x) = 0. (The x-axis is also an asymptote).

$\{x \in \mathbb{R}: x \neq 1\}$ is the domain
$\{f(x) \in \mathbb{R}: f(x) \neq 0\}$ is the range

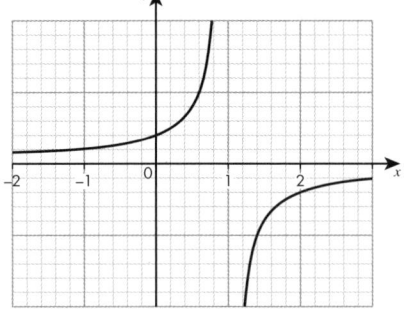

TARGETING DISTINCTION*

6. The function f is defined by $f(x) = x^2 - 4x + 4$, $\{x \in \mathbb{R}: x \geq 2\}$.
   a) Sketch f(x).
   b) State the range of f(x).

7. The function g is defined by $g(x) = \dfrac{1}{x+2}$, $\{x \in \mathbb{R}: x > -2\}$.
   a) Sketch g(x).
   b) State the range of g(x).

8. h(x) = 2 sin x −1 for values of x between −180° and +180°.
   a) Sketch h(x).
   b) State the domain of h(x).
   c) State the range of h(x).

5.2 Composite functions

THIS SECTION WILL SHOW YOU HOW TO ...
✓ find composite functions
✓ sketch the graphs of composite functions
✓ find the domain and range of composite functions

KEY WORDS
✓ composite function

A **composite function** is a combination of two functions that creates a third function. For two functions f(x) and g(x), the function created by substituting g(x) into f(x) is called fg(x). You work this out by evaluating g(x) first and then substituting your answer into f(x).

EXAMPLE 9

Given that $h(x) = x^2$ and $k(x) = x + 4$, find:

a) kh(3)   **b)** kh(−2)   **c)** hk(5).

a) h means 'square' and k means 'add 4'.
kh(3) means 'start with 3, square it, then add 4'.
So kh(3) = $3^2 + 4$
= 9 + 4
= 13

b) kh(−2) = $(−2)^2 + 4$
= 4 + 4
= 8

c) hk(5) is the other way round. It means 'start with 5, add 4, then square it'.
hk(5) = $(5 + 4)^2$
= 9^2
= 81

EXAMPLE 10

The functions f(x) and g(x) are defined as $f(x) = 5x − 3$ and $g(x) = \frac{1}{2}x + 1$.
Find the value of:

a) f(4)   **b)** fg(4)   **c)** ff(4).

a) f(4) = 5 × 4 − 3
= 20 − 3
= 17

b) g(4) = $\frac{1}{2}$ × 4 + 1
= 2 + 1
= 3
f(3) = 5 × 3 − 3
= 15 − 3
= 12

c) f(4) = 17
f(17) = 5 × 17 − 3
= 85 − 3
= 82

EXAMPLE 11

The functions f(x) and g(x) are defined as $f(x) = 5x − 3$ and $g(x) = \frac{1}{2}x + 1$.
Find an expression (in terms of x) to represent:

a) fg(x)   **b)** gf(x)   **c)** ff(x).

a) Substitute g(x) into f(x).
f(x) = $5(\frac{1}{2}x + 1) − 3$
= $2\frac{1}{2}x + 5 − 3$
= $2\frac{1}{2}x + 2$

b) Substitute f(x) into g(x).
g(x) = $\frac{1}{2}(5x − 3) + 1$
= $2\frac{1}{2}x − 1\frac{1}{2} + 1$
= $2\frac{1}{2}x − \frac{1}{2}$

c) Substitute f(x) into f(x).
f(x) = $5(5x − 3) − 3$
= $25x − 15 − 3$
= $25x − 18$

EXERCISE 5C

TARGETING MERIT

1. $s(x) = x + 4$ and $t(x) = \frac{x}{2}$

   a) Find s(2) and ts(2).
   b) Find s(3) and ts(3).
   c) Find s(6) and ts(6).
   d) Find an expression for ts(x).
   e) Find t(2) and st(2).
   f) Find t(3) and st(3).
   g) Find t(−10) and st(−10).
   h) Find an expression for st(x).

2. $r(x) = \sqrt{x}$ and $a(x) = 2x + 1$.

   a) Find a(0), a(4) and a(12).
   b) Find ra(0), ra(4) and ra(12).
   c) Find an expression for ra(x).

3. Given that $m(x) = 3x$, find

   a) m(2) and mm(2)
   b) m(4) and mm(4)
   c) an expression for mm(x).

4. Given that $f(x) = 3x$ and $g(x) = x − 6$, find an expression for

   a) fg(x)
   b) gf(x).

5. Given that $a(x) = x + 4$ and $b(x) = x − 7$, show that ab(x) and ba(x) are identical.

TARGETING DISTINCTION

6. Given that $f(x) = 2^x$ and $g(x) = 2x − 1$, find

   a) gf(2)
   b) fg(2)
   c) ff(3)
   d) gg(6).

7. $f(x) = 3x + 1$ and $g(x) = 2x − 2$

   a) Find an expression for gf(x). Write your answer as simply as possible.
   b) Find fg(x). Write your answer as simply as possible.

8. In each case, find fg(x). Write your answer as simply as possible.

   a) $f(x) = x^2$ and $g(x) = 3x + 4$
   b) $f(x) = 2x + 3$ and $g(x) = 3x − 4$
   c) $f(x) = \frac{x}{2} + 4$ and $g(x) = 4x − 2$
   d) $f(x) = 12 − x$ and $g(x) = 2x + 8$

9. Given that $h(x) = 10 − x$ and $k(x) = 20 − x$, find

   a) hk(x)
   b) kh(x)
   c) kk(x).

10. Given that $h(x) = x^2$ and $k(x) = \frac{12}{x}$, find

    a) hh(x)
    b) hk(x)
    c) kh(x)
    d) kk(x).

11. $m(x) = x^2 + 2x$ and $n(x) = 2x − 1$

    a) Find mm(2).
    b) Show that $nn(x) = 4x − 3$.
    c) Show that $mn(x) = 4x^2 − 1$.

12. $f(x) = \frac{1}{x-4}$ and $g(x) = 3x + 1$

    a) Find fg(x).
    b) Find gf(x), writing your answer as a single fraction.

13. Given that $f(x) = 2x + 1$ and $g(x) = x^2 − x$, solve the equation gf(x) = 0.

14. Given that $f(x) = x^2 + 2x$ and $g(x) = x − 3$, solve the equation fg(x) = 0.

15. Given that $f(x) = \frac{2x-2}{x-4}$ and $g(x) = \frac{2x-1}{3}$, solve the equation fg(x) = −4.

5.3 Inverse functions

THIS SECTION WILL SHOW YOU HOW TO ...
✓ find inverse functions

KEY WORDS
✓ inverse

The **inverse** of a function f(x) is the function that has the opposite, or reverse, effect to the original and is written as $f^{-1}(x)$.

The function $f(x) = 4x - 2$ 'multiplies x by 4 and subtracts 2'.

The inverse function will 'add 2 and then divide by 4'. $f^{-1}(x) = \frac{x+2}{4}$.

To work out the inverse function, write $y = f(x)$, so $y = 4x - 2$, and then rearrange the equation to make x the subject.

$y + 2 = 4x \Rightarrow \frac{y+2}{4} = x$

Finally replace y with x to give $f^{-1}(x) = \frac{x+2}{4}$

Only one-to-one functions or many-to-one functions where the domain has been restricted have inverses.

EXAMPLE 12

$f(x) = (x + 2)^2 - 3$ for $x \geq 1$

a) Find an expression for $f^{-1}(x)$. **b)** Find $f^{-1}(13)$. **c)** Solve the equation $f^{-1}(x) = 8$.

a)
$y = (x + 2)^2 - 3$
$y + 3 = (x + 2)^2$
$\sqrt{y + 3} = x + 2$
$\sqrt{y + 3} - 2 = x$
$\sqrt{x + 3} - 2 = f^{-1}(x)$

b)
$f^{-1}(x) = \sqrt{x + 3} - 2$
$f^{-1}(13) = \sqrt{13 + 3} - 2$
$f^{-1}(13) = \sqrt{16} - 2$
$f^{-1}(13) = 4 - 2$
$f^{-1}(13) = 2$

c)
$\sqrt{x + 3} - 2 = 8$
$\sqrt{x + 3} = 10$
$(\sqrt{x + 3})^2 = 100$
$x + 3 = 100$
$x = 97$

EXAMPLE 13

Given that $f(x) = (x + 1)^2 + 3$ for $x \geq -1$, find

a) an expression for $f^{-1}(x)$ **b)** the value of x when $f^{-1}(x) = f(-1)$.

a)
$y = (x + 1)^2 + 3$
$y - 3 = (x + 1)^2$
$\sqrt{y - 3} = x + 1$
$f^{-1}(x) = \sqrt{x - 3} - 1$

b) $f(-1) = (-1 + 1)^2 + 3$
$f(-1) = 3$
$3 = \sqrt{x - 3} - 1$
$4 = \sqrt{x - 3}$
$16 = x - 3$
$19 = x$
$f^{-1}(x) = f(-1)$ when $x = 19$

EXAMPLE 14

Given that $f(x) = \frac{2x-1}{x+3}$ for $\{x \in \mathbb{R}: x > -3\}$, find $f^{-1}(x)$ and the domain and range of this inverse function.

Write $y = f(x)$:
$$y = \frac{2x-1}{x+3}$$

Rearrange the equation to make x the subject
$$xy + 3y = 2x - 1$$
$$3y + 1 = 2x - xy$$

Factorise by taking x outside a bracket:
$$3y + 1 = x(2 - y)$$
$$x = \frac{3y+1}{2-y}$$

Finally replace y with x to give
$$f^{-1}(x) = \frac{3x+1}{2-x}$$

The domain is $\{x \in \mathbb{R}: x < 2\}$.

The range is the domain of the original function $\{f^{-1}(x) \in \mathbb{R}: f^{-1}(x) > -3\}$.

HINTS AND TIPS

The domain of an inverse function is the same as the range of the original function and the range of the inverse of a function is the same as the domain of the original function.
An inverse function reverses the effect of the original function (it takes elements in the range of the original and maps them back to the domain).

EXAMPLE 15

Given that $g(x) = \sqrt{x-5}$, $\{x \in \mathbb{R}: x > 5\}$,

a) state the range of $g(x)$
b) find $g^{-1}(x)$ and state its domain and range
c) sketch $g(x)$ and $g^{-1}(x)$ and the line $y = x$ on the same axes.
d) describe the transformation that maps $g(x)$ to $g^{-1}(x)$.

a) The range is $\{g(x) \in \mathbb{R}: g(x) > 0\}$.

b) $g^{-1}(x) = x^2 + 5$

The domain is $\{x \in \mathbb{R}: x > 0\}$ and the range is $\{g^{-1}(x) \in \mathbb{R}: g^{-1}(x) > 5\}$.

c)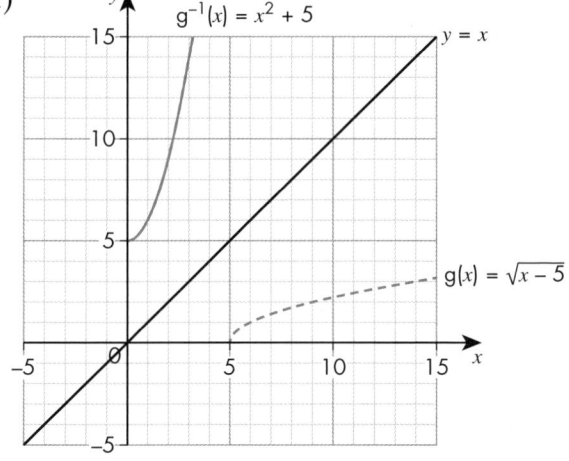

d) A reflection in the line $y = x$.

EXERCISE 5D

TARGETING MERIT

1. Find $f^{-1}(x)$ for each of these functions.

a) $f(x) = x + 7$ b) $f(x) = 8x$ c) $f(x) = \frac{x}{5}$ d) $f(x) = x - 3$

2. Find $f^{-1}(x)$ for each of these functions.
   a) $f(x) = \frac{x}{3} - 2$
   b) $f(x) = 4(x - 5)$
   c) $f(x) = \frac{x+4}{5}$
   d) $f(x) = \frac{3x-6}{2}$
   e) $f(x) = 3\left(\frac{x}{2} + 4\right)$
   f) $f(x) = 4x^3$

3. Find $f^{-1}(x)$ for each of these functions.
   a) $f(x) = \frac{8}{x}$, $\{x \in \mathbb{R}: x \neq 0\}$,
   b) $f(x) = \frac{20}{x} - 1$, $\{x \in \mathbb{R}: x \neq 0\}$,
   c) $f(x) = \frac{2}{x+1}$, $\{x \in \mathbb{R}: x \neq -1\}$

4. Work out $f^{-1}(x)$ for each of these functions.
   a) $f(x) = (x - 2)^2 + 4$
   b) $f(x) = \frac{3x+1}{4}$
   c) $f(x) = \frac{8-3x}{5x}$, $\{x \in \mathbb{R}: x \neq 0\}$,
   d) $f(x) = \frac{5}{9}(x - 32)$
   e) $f(x) = (x - 2)^3$

TARGETING DISTINCTION

5. Given that $f(x) = \frac{2}{x}$ $\{x \in \mathbb{R}: x \neq 0\}$, find $f^{-1}(x)$ and explain what you notice.

6. Given that $g(x) = \frac{5x+4}{x-3}$ $\{x \in \mathbb{R}: x \neq 3\}$, explain what happens to $g(x)$ as x approaches 3.

7. Given that $h(x) = x^3 - 2$, find $h^{-1}(x)$ and sketch both graphs on the same axes.

TARGETING DISTINCTION*

8. Given that $g(x) = \sqrt{x - 6}$, $\{x \in \mathbb{R}: x > 6\}$,
   a) state the range of $g(x)$
   b) find $g^{-1}(x)$ and state the domain and range
   c) sketch $g(x)$ and $g^{-1}(x)$ and the line $y = x$ on the same axes
   d) describe the transformation that maps $g(x)$ to $g^{-1}(x)$.

9. Given that $f(x) = \frac{\sqrt{x}-5}{3}$ for $\{x \in \mathbb{R}: x \geq 0\}$, find $f^{-1}(2x)$ and state the domain and range of this inverse function.

10. Given that $f(x) = \frac{4x-3}{x-5}$ for $\{x \in \mathbb{R}: x > 5\}$, find $f^{-1}(x)$ and state the domain and range of the inverse function.

5.4 Transforming functions

THIS SECTION WILL SHOW YOU HOW TO ...
✓ understand the effects that transformations have on functions and their equations

KEY WORDS
✓ transformed
✓ algebraic equation

When a function is **transformed**, the **algebraic equation** will change.
The rules for transforming functions are the same as for transforming graphs (see Chapter 4).
The effects on the coordinates of a function after transformations are as follows:

- f(x) + a is the translation with vector $\begin{pmatrix} 0 \\ a \end{pmatrix}$; so ($x$, y) becomes (x, $y + a$).
- f($x + a$) is the translation $\begin{pmatrix} -a \\ 0 \end{pmatrix}$; so ($x$, y) becomes ($x - a$, y).
- −f(x) is a reflection in the x axis; so (x, y) becomes (x, $-y$).
- f($-x$) is a reflection in the y-axis; so (x, y) becomes ($-x$, y).
- af(x) is a stretch in the y-direction, scale factor a; so (x, y) becomes (x, ay).
- f(ax) is a stretch in the x-direction, scale factor $\frac{1}{a}$; so (x, y) becomes ($\frac{x}{a}$, y).

To find the algebraic equation of a function after these transformations:

- f(x) + a add a to the equation
- f($x + a$)   substitute ($x + a$) in for x
- −f(x)   multiply the equation by −1
- f($-x$)   substitute − x for x
- af(x)   multiply the equation by a
- f(ax)   substitute ax in for x

HINTS AND TIPS

Note that f($x + a$) is a translation in the *negative* x-direction.

Some questions can involve combinations of transformations. For these, follow the normal rules of the order of operations (brackets, indices, division, multiplication, addition, subtraction) to find the new equation of the function.

EXAMPLE 16

f(x) = $x^2 + 3x - 4$

Write the algebraic equation for the following, describing each transformation.

a) $y = $ f(x) + 5   b) $y = $ f($x - 2$)   c) $y = -$f(x)

d) $y = $ f($-x$)   e) $y = 2$f(x)   f) $y = $ f($3x$)

a) $f(x) + 5 = x^2 + 3x - 4 + 5$
   $y = x^2 + 3x + 1$
   translation with vector $\begin{pmatrix} 0 \\ 5 \end{pmatrix}$

b) $f(x - 2) = (x - 2)^2 + 3(x - 2) - 4$
   $y = x^2 - 4x + 4 + 3x - 6 - 4$
   $y = x^2 - x - 6$
   translation with vector $\begin{pmatrix} 2 \\ 0 \end{pmatrix}$

c) $-f(x) = -(x^2 + 3x - 4)$
   $y = -x^2 - 3x + 4$
   reflection in the line $y = 0$ (x-axis)

d) $f(-x) = (-x)^2 + 3(-x) - 4$
   $y = x^2 - 3x - 4$
   reflection in the line $x = 0$ (y-axis)

e) $2f(x) = 2(x^2 + 3x - 4)$
   $y = 2x^2 + 6x - 8$
   stretch, scale factor 2 parallel to the y-axis

f) $f(3x) = (3x)^2 + 3(3x) - 4$
   $y = 9x^2 + 9x - 4$
   stretch, scale factor $\frac{1}{3}$ parallel to the x-axis

EXAMPLE 17

$f(x) = x^2(x + 1)$

Write the algebraic equation for
a) $y = 2f(x) - 3$ b) $y = -f(2x + 5)$

a) $y = 2(x^2(x + 1)) - 3$
   $y = 2x^3 + 2x^2 - 3$

b) $y = -(2x + 5)^2((2x + 5) + 1)$
   $y = -(4x^2 + 20x + 20)(2x + 6)$

EXERCISE 5E

TARGETING MERIT

1. $f(x) = 3x - 2$
   Write the algebraic equation for
   a) $y = 3f(x)$ b) $y = f(x) + 7$ c) $y = f(2x)$ d) $y = f(x - 2)$

2. Describe the transformation that maps $g(x)$ to $g(3x)$.

3. Describe the transformation that maps $h(x)$ to $h(-x)$.

TARGETING DISTINCTION

4. The diagram shows the graphs of f(x), g(x) and h(x).

f(x) has been transformed to make both g(x) and h(x).

a) Write the equation of f(x).

b) Write the equation of g(x) in terms of f(x).

c) Write the equation of h(x) in terms of f(x).

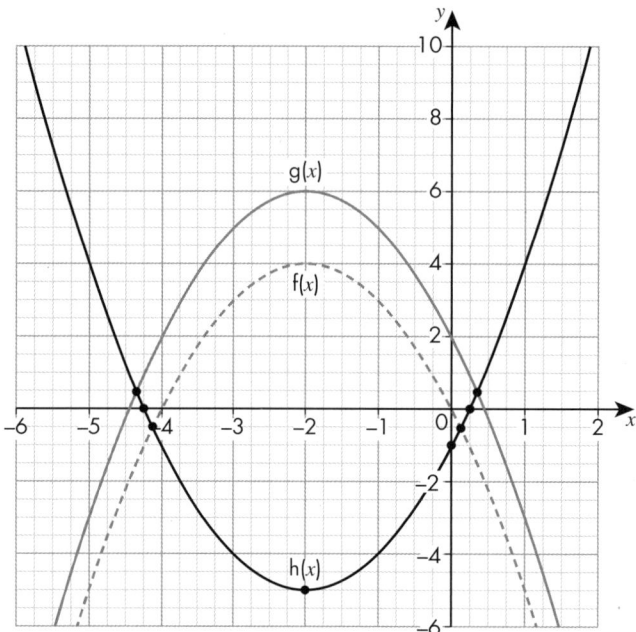

5. The function $y = f(x)$ is transformed by a translation by $\begin{pmatrix} 4 \\ -1 \end{pmatrix}$.

Write the equation of the new function in terms of f(x).

6. The function $y = g(x)$ is transformed by a stretch in the x-direction with scale factor 3 followed by a stretch in the y-direction with scale factor 2.

Write the equation of the new function in terms of g(x).

TARGETING DISTINCTION*

7. $f(x) = \dfrac{x^2 + 1}{7}$

Write the algebraic equation for $y = f(-3x) - 2$, writing your answer as a single fraction.

8. The function f(x) has been transformed to make graph C.

a) Write the equation of f(x).

b) Write the equation of graph C.

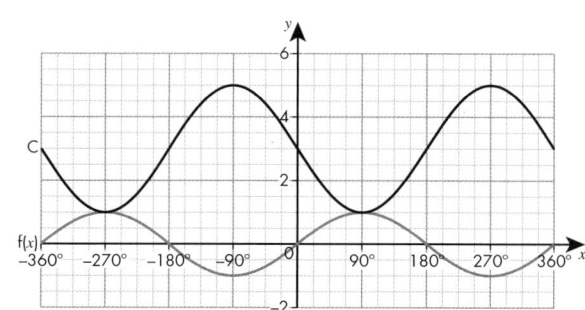

Exam-style questions

1. f(n) = n(n + 1) where n is an integer.
   Find f(10) and f(−10). [2 marks]

2. State whether the following mappings are one-to-one, many-to, one, one-to-many or many-to-many.

   a) i)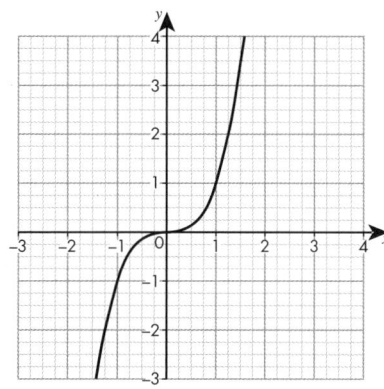

   ii)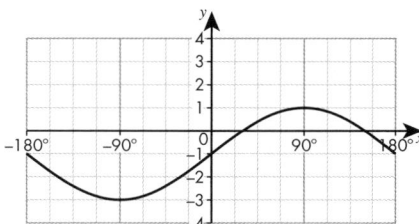

   iii)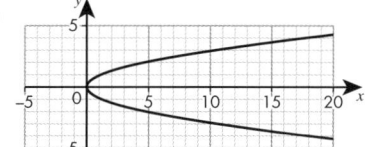

   iv)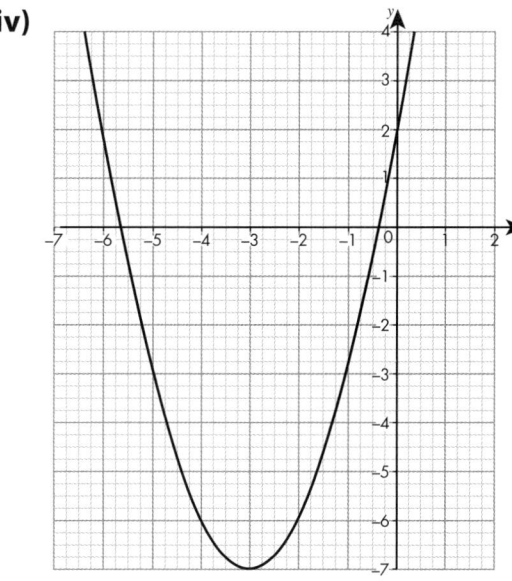

   [4 marks]

   b) Which one of these mappings is not a function? [1 mark]

3. $f(x) = 2x + x^2$ and $x \geq 3$.

   a) Find f(3.5). [2 marks]

   b) Write down the range of f(x). [1 mark]

4. $g(x) = \frac{x+2}{x+4}$ for all possible values of x.

   a) What value of x must be excluded from the domain? [1 mark]

   b) Find the inverse function of g(x). [3 marks]

5. A function f(x) is defined as [3 marks]

$f(x) = x^2 + 1 \qquad 0 \leq x < 1$

$f(x) = 2x \qquad 1 \leq x < 2$

$f(x) = 4 \qquad 2 \leq x < 3$

Sketch the graph of f(x) for values of x from 0 to 3.

6. Describe, in order, the sequence of transformations of the graph $y = f(x)$ given by the following equations.

a) $y = 3f(x - 1) - 2$ [3 marks]

b) $y = -f\left(\dfrac{1}{2}x\right) + 3$ [3 marks]

7. Here is the graph of $y = f(x)$, where $f(x) = (x + 2)^2$.

f(x) is transformed to graph of $y = g(x)$, where $g(x) = -f(x - 1) + 2$.

Write the equation of g(x).

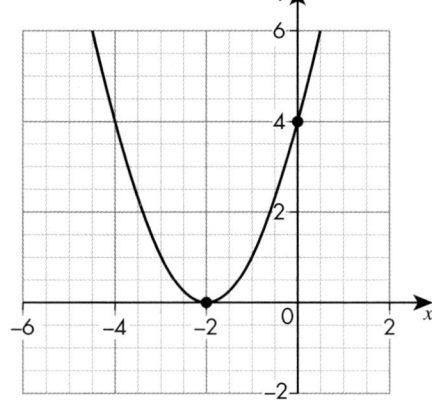

8. The function f is defined by $f(x) = x^2 + 6x + 2$.

a) Sketch f(x) and define its range. [4 marks]

b) Explain why the function f(x) does not have an inverse. [1 mark]

9. The function f is defined by $f(x) = \dfrac{1}{x-2} \{x \in \mathbb{R} : x \neq 2\}$.

Write down the equation of ff(x), writing your answer as a fraction in its simplest form. [3 marks]

6 Equations and inequalities

6.1 Solve equations

THIS SECTION WILL SHOW YOU HOW TO ...
✓ solve equations involving algebraic fractions
✓ solve equations involving trigonometric functions

KEY WORDS
✓ algebraic fractions
✓ trigonometric equations

Equations involving **algebraic fractions** will require several steps to reach a solution.

To solve these types of equations, first write the algebraic fractions over one numerator, or find a common denominator, and multiply though by this value (or expression).

EXAMPLE 1

Solve the equation $\frac{x+1}{3} - \frac{x-3}{2} = x - 4$.

Method 1

Write the expression on the left as a single fraction:

$$\frac{2(x+1) - 3(x-3)}{3 \times 2} = x - 4$$

Multiply by 6: $\quad 2(x+1) - 3(x-3) = 6(x-4)$
Remove the brackets: $\quad 2x + 2 - 3x + 9 = 6x - 24$
$\quad\quad\quad\quad\quad\quad\quad\quad\quad 11 - x = 6x - 24$
Rearrange: $\quad\quad\quad\quad\quad\quad 35 = 7x \Rightarrow x = 5$

Method 2

The lowest common multiple (LCM) of 2 and 3 is 6, so multiply every term by 6.

$6 \times \frac{x+1}{3} - 6 \times \frac{x-3}{2} = 6(x-4) \Rightarrow 2(x+1) - 3(x-3)$
$\quad\quad\quad\quad\quad\quad\quad\quad\quad\quad = 6(x-4)$

This clears the fractions. Now continue as before.

HINTS AND TIPS

Remember to apply negatives before brackets to all the signs in the bracket.

EXAMPLE 2

Solve

a) $\frac{3}{x} - \frac{5}{3x} = 2$
b) $\frac{x+1}{x+2} \div \frac{x^2-1}{2x^2+x-6} = \frac{3}{4}$

a) Rewrite the equation to form a common denominator (multiply 3 by 3 to make equivalent fractions):

$$\frac{3}{x} - \frac{5}{3x} = 2$$

$$\frac{3(3) - 5}{3x} = 2$$

Multiply both sides by $3x$: $\quad 9 - 5 = 6x$

Solve: $\quad\quad\quad\quad\quad\quad\quad 4 = 6x$

$$x = \frac{4}{6} = \frac{2}{3}$$

b) This equation looks more difficult than it is. By factorising you can simplify the fractions to make it much easier to then solve.

Rewrite the equation to form a multiplication (invert the second fraction) and factorise the quadratic expressions.

$$\frac{x+1}{x+2} \div \frac{x^2-1}{2x^2+x-6} = \frac{3}{4}$$

$$\frac{x+1}{x+2} \times \frac{(2x-3)(x+2)}{(x+1)(x-1)} = \frac{3}{4}$$

Cancel all the common factors: $\quad \frac{(2x-3)}{(x-1)} = \frac{3}{4}$

Cross multiply: $\quad\quad\quad\quad\quad 4(2x-3) = 3(x-1)$

Expand brackets: $\quad\quad\quad\quad 8x - 12 = 3x - 3$

Solve for x: $\quad\quad\quad\quad\quad\quad 5x = 9$

$$x = \frac{9}{5}$$

Solving **trigonometric** equations is similar to solving algebraic equations.

EXAMPLE 3

Find all the solutions to $2 \sin(x - 30°) - 1 = 0$ for $0° \leq x \leq 180°$.

First add 1 to both sides of the equation: $\quad 2\sin(x - 30°) = 1$

Then divide by 2: $\quad\quad\quad\quad\quad\quad\quad\quad \sin(x - 30°) = \frac{1}{2}$

As $\sin^{-1}\left(\frac{1}{2}\right) = 30°$, this tells you that $(x - 30°) = 30°$ between $0° \leq x \leq 180°$.

Let $(x - 30°) = X$. Adjust the interval by subtracting 30° from each end to get a new interval of $-30° \leq X \leq 150°$.

Solving $\sin X = \frac{1}{2}$ gives $X = 30°$

You can find all the possible solutions by considering the graph of $\sin X$ between $-30°$ and $150°$.

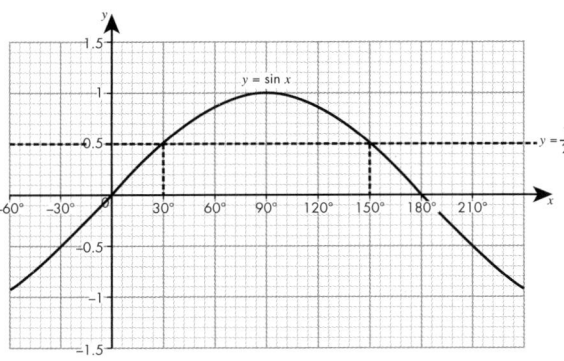

Solutions of $\sin X = \frac{1}{2}$ are $30°$ and $150°$.

So, as $X = (x - 30°)$, then $(x - 30°) = 30°$ or $(x - 30°) = 150°$.

This gives solutions of $x = 60°$ or $x = 180°$.

These solutions are both within $0° \leq x \leq 180°$ and so are both valid.

Always check your solutions by substituting them back into the original equation:

$2 \sin (60° - 30°) - 1 = 0 \Rightarrow 2 \sin 30° - 1 = 0 \Rightarrow \sin 30° = \frac{1}{2}$, which is correct

$2 \sin (180° - 30°) - 1 = 0 \Rightarrow 2 \sin 150° - 1 = 0 \Rightarrow \sin 150° = \frac{1}{2}$, which is correct

EXERCISE 6A

TARGETING PASS

1. Solve these equations.

   a) $\frac{3x + 5}{2} = x + 6$

   b) $\frac{3x + 1}{2} = \frac{9x - 5}{5}$

   c) $10 - x = \frac{18 - 3x}{2}$

2. Solve these equations.

   a) $\frac{x + 1}{2} + \frac{x + 2}{5} = 3$

   b) $\frac{x + 2}{4} + \frac{x + 1}{7} = 3$

   c) $\frac{4x + 1}{3} - \frac{x + 2}{4} = 2$

3. Solve these equations.

   a) $\frac{x}{3} + \frac{x}{4} = \frac{x + 1}{2}$

   b) $\frac{12 - x}{2} = \frac{11 - x}{3}$

   c) $\frac{x + 1}{4} + \frac{x + 2}{3} = 12 - x$

4. The perimeter of this triangle is 18.

   Calculate the length of each side.

   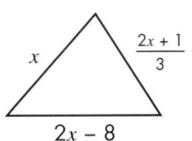

5. The angles of a triangle are x, $\frac{2x + 10}{2}$ and $\frac{3x}{2}$ degrees.

   Use the fact that the angles of a triangle add up to 180 degrees to calculate the size of each angle.

   HINTS AND TIPS

   Write every step of your working clearly and ensure that when you multiply to simplify your equation (to eliminate the fractions) you multiply all values on both sides of the equation.

TARGETING MERIT

6. a) $\dfrac{3}{x} - \dfrac{3x-1}{2x} = -5$ b) $\dfrac{7}{2-x} - \dfrac{4}{2-x} = 2$

   c) $\dfrac{x^2 - x - 12}{x^2 - 16} = 7$ d) $\dfrac{2x^2 + x - 3}{x^2 + 4x - 5} = -2$

   e) $\dfrac{6x^2 - 7x - 5}{2x^2 - 5x - 3} = \dfrac{2}{3}$

   f) $\dfrac{x^2 - 25}{3x^2 + 7x - 6} \div \dfrac{x^2 - 4x - 5}{3x^2 + x - 2} = -3$

HINTS AND TIPS

For question 6 parts **c** to **f**, first factorise the expressions, then simplify and solve.

TARGETING DISTINCTION

7. Solve the following equations in the interval $0° \leq x \leq 360°$.

   a) $\sin x = \dfrac{\sqrt{3}}{2}$ b) $2\tan x = 2$

   c) $\cos x = \dfrac{\sqrt{2}}{2}$ d) $3\cos x = \dfrac{3}{2}$

HINTS AND TIPS

When solving trigonometric equations, leave your answers in surd form when appropriate. You can use your calculator to check your answers.

TARGETING DISTINCTION*

8. Solve the following equations in the interval $-180° \leq x \leq 180°$.

   a) $2\sin(x - 60°) = 1$ b) $2\cos(x + 45) = \sqrt{3}$

   c) $\tan 3x - \sqrt{3} = 0$ d) $2\sin 2x - \sqrt{2} = 0$

6.2 Solve quadratic equations

THIS SECTION WILL SHOW YOU HOW TO ...

✓ solve quadratic equations by factorising
✓ solve quadratic equations by using the quadratic formula
✓ solve quadratic equations by completing the square

KEY WORDS

✓ factorising
✓ completing the square
✓ real solutions
✓ factors
✓ coefficient
✓ discriminant
✓ quadratic formula
✓ constant

Solving the general quadratic equation by factorising

The general form of a quadratic equation is $ax^2 + bx + c = 0$ where a, b and c are positive or negative whole numbers. Before you can solve any quadratic equation by **factorising**, you must rearrange it to this form.

It is easier to solve a quadratic equation if a is positive. If a is negative and the equation is equal to zero, then you can multiply both sides by -1 to make a positive.

EXAMPLE 4

Solve $x^2 - 6x + 9 = 0$.

This factorises to give $(x - 3)(x - 3) = 0$.
Which can be written as $(x - 3)^2 = 0$.
Hence, there is only one solution, $x = 3$.

EXAMPLE 5

Solve the quadratic equation $12x^2 - 28x = -15$.

First, rearrange the equation to the general form: $12x^2 - 28x + 15 = 0$

Factorise: $(2x - 3)(6x - 5) = 0$

The only way the product can equal 0 is if the value of the expression in one of the brackets is 0.

Hence: either $2x - 3 = 0$ or $6x - 5 = 0$
$\Rightarrow \qquad 2x = 3$ or $6x = 5$
$\Rightarrow \qquad x = \frac{3}{2}$ or $x = \frac{5}{6}$

So the solution is $x = 1\frac{1}{2}$ or $x = \frac{5}{6}$

Note: It is more accurate to give your answer as a fraction when it is appropriate. A solution that is rounded is less accurate.

Special cases

Sometimes the value of b and/or c is zero. (Note that if a is zero, then the equation is no longer a quadratic equation.)

EXAMPLE 6

Solve these quadratic equations.   **a)** $3x^2 - 4 = 0$ **b)** $6x^2 - x = 0$

a) Rearrange: $\qquad\qquad\qquad 3x^2 = 4$

Divide both sides by 3: $\qquad x^2 = \frac{4}{3}$

Take the square root on both sides: $x = \pm\sqrt{\frac{4}{3}} = \pm\frac{2}{\sqrt{3}} = \pm\frac{2\sqrt{3}}{3}$

Note: A square root can be positive or negative. The symbol $\pm$ indicates that the square root has a positive and a negative value, so you need to give *both* as the values of x.

b) There is a common factor of x, so factorise as $x(6x - 1) = 0$.

There is only one set of brackets this time but each factor can be equal to zero, so $x = 0$ or $6x - 1 = 0$

Therefore, $x = \frac{1}{6}$

EXAMPLE 7

Solve this equation.

$\frac{3}{x-1} - \frac{2}{x+1} = 1$

Use the rule for combining fractions and cross-multiply to take the denominator of the left-hand side to the right-hand side. Use brackets to help with expanding and to avoid problems with minus signs.

$$3(x + 1) - 2(x - 1) = (x - 1)(x + 1)$$

The right-hand side is the difference of two squares:

$$3x + 3 - 2x + 2 = x^2 - 1$$

Rearrange into the general quadratic form: $\quad x^2 - x - 6 = 0$

Factorise and solve: $\quad (x - 3)(x + 2) = 0$

$$x = 3 \text{ or } x = -2$$

EXERCISE 6B

TARGETING PASS

Give your answers either in rational form or as mixed numbers.

1. Solve these equations.
    a) $3x^2 + 8x - 3 = 0$
    b) $6x^2 - 5x - 4 = 0$
    c) $4t^2 - 4t - 35 = 0$
    d) $18t^2 + 9t + 1 = 0$
    e) $12x^2 - 16x - 35 = 0$
    f) $15t^2 + 4t - 35 = 0$
    g) $28x^2 - 85x + 63 = 0$
    h) $16t^2 - 1 = 0$
    i) $4x^2 + 9x = 0$
    j) $9m^2 - 24m - 9 = 0$

2. Rearrange these equations into the general form and then solve them.
    a) $x^2 - x = 12$
    b) $8x(x + 1) = 30$
    c) $(x + 1)(x - 2) = 40$
    d) $13x^2 = 11 - 2x$
    e) $8x^2 + 6x + 3 = 2x^2 + x + 2$
    f) $25x^2 = 10 - 45x$
    g) $8x - 16 - x^2 = 0$
    h) $(2x + 1)(5x + 2) = (2x - 2)(x - 2)$

3. Here are three equations.

    A: $(x - 1)^2 = 0$ \quad B: $3x + 2 = 5$ \quad C: $x^2 - 4x = 5$

    a) Give one mathematical fact that equations A and B have in common.

    b) Give a mathematical reason why equation B is different from equations A and C.

4. Pythagoras' theorem states that the sum of the squares of the two shorter sides of a right-angled triangle is equal to the square of the longest side (hypotenuse).

   A right-angled triangle has sides of length $5x - 1$, $2x + 3$ and $x + 1$ cm.

   a) Show that $20x^2 - 24x - 9 = 0$.

   b) Find the area of the triangle.

5. Show that each algebraic fraction simplifies to the given expression.

   a) $\dfrac{2}{x+1} + \dfrac{5}{x+2} = 3$ simplifies to $3x^2 + 2x - 3 = 0$

   b) $\dfrac{4}{x-2} + \dfrac{7}{x+1} = 3$ simplifies to $3x^2 - 14x + 4 = 0$

   c) $\dfrac{3}{2x-1} - \dfrac{4}{3x-1} = 1$ simplifies to $x^2 - x = 0$

6. Solve these equations.

   a) $\dfrac{4}{x+1} + \dfrac{5}{x+2} = 2$ b) $\dfrac{18}{4x-1} - \dfrac{1}{x+1} = 1$ c) $\dfrac{2x-1}{2} - \dfrac{6}{x+1} = 1$

7. a) Solve the equation $x^2 - 7x + 10 = 0$.

   b) Use your answer to part **a** to solve the equation $x - 7\sqrt{x} + 10 = 0$.

Solving quadratic equations using the quadratic formula

Many quadratic equations cannot be solved by factorising because they do not have simple factors. For example, try to factorise $x^2 - 4x - 3 = 0$ or $3x^2 - 6x + 2 = 0$. You will find that it is impossible.

One way to solve this type of equation is to use the **quadratic formula**. This formula can be used to solve *any* quadratic equation that has real solutions.

The solutions of the equation $ax^2 + bx + c = 0$ is given by the formula:

$$x = \dfrac{-b \pm \sqrt{b^2 - 4ac}}{2a}$$

where *a* and *b* are the **coefficients** of x^2 and x, respectively, and *c* is the **constant** term.

Real solutions

$b^2 - 4ac$ is called the **discriminant** of a quadratic equation in the form $ax^2 + bx + c$.

If $b^2 - 4ac > 0$, then the quadratic equation will have two **real solutions**.

If $b^2 - 4ac = 0$, then the quadratic equation will have one real solution.

If $b^2 - 4ac < 0$, then the quadratic equation will have no real solutions.

EXAMPLE 8

Solve $5x^2 - 11x - 4 = 0$, giving solutions correct to 2 decimal places.

Write the quadratic formula: $x = \dfrac{-b \pm \sqrt{b^2 - 4ac}}{2a}$

Substitute $a = 5$, $b = -11$ and $c = -4$:

$$x = \dfrac{(11) \pm \sqrt{(-11)^2 - 4(5)(-4)}}{2(5)}$$

$$x = \dfrac{11 \pm \sqrt{121 + 80}}{10} = \dfrac{11 \pm \sqrt{201}}{10} \Rightarrow x = 2.52 \text{ or } -0.32$$

HINTS AND TIPS

Note that the values for a, b and c have been put into the formula in brackets.

This is to avoid mistakes in the calculation. It is a very common mistake to get the sign of b wrong or to think that -11^2 is -121. Using brackets will help you do the calculation correctly.

HINTS AND TIPS

If you are asked in an examination question to solve a quadratic equation to one or two decimal places, you can be confident that it can be solved by the quadratic formula and cannot be factorised.

EXERCISE 6C

TARGETING PASS

Use the quadratic formula to solve the equations in questions **1–6**. Give your answers to 2 decimal places.

1. $2x^2 + x - 8 = 0$
2. $3x^2 + 5x + 1 = 0$
3. $x^2 - x - 10 = 0$
4. $4x^2 - 9x + 4 = 0$
5. $7x^2 + 3x - 2 = 0$
6. $5x^2 - 10x + 1 = 0$

7. A rectangular lawn is 2 m longer than it is wide.

   The area of the lawn is 21 m². A gardener is going to put edging strips around the lawn, which are sold in lengths of $1\tfrac{1}{2}$ m. How many strips will they need to buy?

8. Shaun is solving a quadratic equation, using the formula.

   He correctly substitutes values for a, b and c to get:

   $x = \dfrac{3 \pm \sqrt{37}}{2}$

   What is the equation that Shaun is trying to solve?

9. Explain why the equation $2x^2 + 4x + 3 = 0$ has no real solutions.

10. Work out the number of real solutions for each of these equations.

    a) $x^2 - 10x + 25 = 0$
    b) $2x^2 + x + 1 = 0$
    c) $3x^2 - 7x - 2 = 0$

Solve quadratic equations by completing the square

Another method for solving quadratic equations is **completing the square**. This method can be used as an alternative to the quadratic formula. This method has already been covered in Section 2.4.

To recap:
Remember that $(x+a)^2 = x^2 + 2ax + a^2$
which can be rearranged to give $x^2 + 2ax = (x+a)^2 - a^2$
This is the principle behind completing the square.
There are three basic steps in rewriting $x^2 + px + q$ in the form $(x+a)^2 + b$.

Step 1: Ignore q and just look at the first two terms, $x^2 + px$.

Step 2: Rewrite $x^2 + px$ as $\left(x + \frac{p}{2}\right)^2 - \left(\frac{p}{2}\right)^2$.

Step 3: Bring q back to get $x^2 + px + q = \left(x + \frac{p}{2}\right)^2 - \left(\frac{p}{2}\right)^2 + q$.

EXAMPLE 9

Rewrite $x^2 + 4x - 7$ in the form $(x+a)^2 - b$.
Hence solve the equation $x^2 + 4x - 7 = 0$, giving your answers to 2 decimal places.

Note that $\quad x^2 + 4x = (x+2)^2 - 4$
So $\quad x^2 + 4x - 7 = (x+2)^2 - 4 - 7$
$\quad\quad\quad\quad\quad\quad = (x+2)^2 - 11$

So, to solve $x^2 + 4x - 7 = 0$ by completing the square,
rewrite the equation as $\quad (x+2)^2 - 11 = 0$
then rearrange: $\quad (x+2)^2 = 11$

Taking the square root of both sides gives $\quad x + 2 = \pm\sqrt{11} \Rightarrow x = -2 \pm \sqrt{11}$.

This answer could be left like this, but you are asked to calculate it to 2 decimal places so $x = 1.32$ or $x = -5.32$ (to 2 decimal places)
To solve $ax^2 + bx + c = 0$ when a is not 1, start by dividing through by a.

EXAMPLE 10

Solve by completing the square: $\quad 2x^2 - 6x - 7 = 0$
Divide by 2: $\quad x^2 - 3x - 3.5 = 0$
Complete the square for the first
two terms: $\quad x^2 - 3x = (x - 1.5)^2 - 2.25$
So: $\quad x^2 - 3x - 3.5 = (x - 1.5)^2 - 2.25 - 3.5$
$\quad\quad\quad\quad\quad\quad = (x - 1.5)^2 - 5.75$
The equation becomes: $\quad (x - 1.5)^2 - 5.75 = 0$
Rearrange: $\quad (x - 1.5)^2 = 5.75$
Take the square root of both sides: $\quad x - 1.5 = \pm\sqrt{5.75}$
Write the solution: $\quad x = 1.5 \pm \sqrt{5.75}$

EXAMPLE 11

a) Solve the following equation by completing the square, leaving your answer in surd form.

$-3x^2 + 6x + 5 = 0$

b) Sketch the graph of $y = -3x^2 + 6x + 5$ labelling the minimum point and any points where the graph intersects the coordinate axes.

a) When the coefficient of x^2 is negative, begin by dividing every term by -1. This will produce the correct solutions but remember that this operation will result in reflecting the graph of the function in the x-axis ($-f(x)$).

Divide by -3: $\qquad x^2 - 2x - \frac{5}{3} = 0$

Complete the square for the first two terms: $\qquad x^2 - 2x = (x - 1)^2 - 1$

So: $\qquad x^2 - 2x - \frac{5}{3} = (x - 1)^2 - 1 - \frac{5}{3}$

$\qquad\qquad = (x - 1)^2 - \frac{8}{3}$

The original equation $\qquad x^2 - 2x - \frac{5}{3} = 0$

becomes $\qquad (x - 1)^2 - \frac{8}{3} = 0$

Rearrange: $\qquad (x - 1)^2 = \frac{8}{3}$

$\qquad x - 1 = \pm\sqrt{\frac{8}{3}}$

$\qquad x = 1 \pm \sqrt{\frac{8}{3}}$

$\qquad x = 1 \pm \frac{2\sqrt{6}}{3}$

These solutions will be the points where the graph crosses the x-axis.

b) When x^2 is negative, the quadratic graph will have a maximum point.

$-3x^2 + 6x + 5 = -3(x^2 - 2x) + 5$
$\qquad\qquad\qquad = -3\left((x - 1)^2 - 1\right) + 5$
$\qquad\qquad\qquad = -3(x - 1)^2 + 8$

The maximum point is $(1, 8)$.

The roots are $x = 1 \pm \frac{2\sqrt{6}}{3}$, which give the x-intercepts.

The y-intercept can be found by substituting $x = 0$ into the original equation.

This gives $y = -3 \times 0^2 + 6 \times 0 + 5 = 5$

The y-intercept is $(0, 5)$.

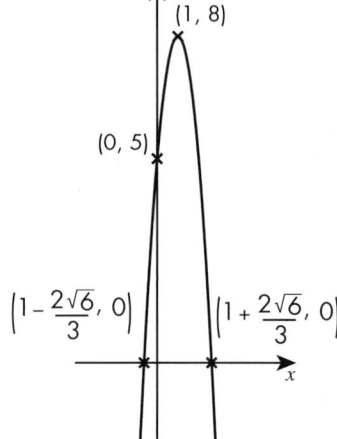

EXERCISE 6D

TARGETING MERIT

1. Solve each equation by completing the square. Give your answer in the form $a + \sqrt{b}$.
   a) $x^2 + 4x - 1 = 0$
   b) $x^2 + 14x - 5 = 0$
   c) $x^2 - 6x + 3 = 0$
   d) $x^2 + 2x - 1 = 0$
   e) $x^2 - 2x - 7 = 0$
   f) $x^2 + 2x - 9 = 0$

2. Solve by completing the square. Give your answers to 2 decimal places.
   a) $x^2 + 2x - 5 = 0$
   b) $x^2 - 4x - 7 = 0$
   c) $x^2 + 2x - 9 = 0$

3. Jorge writes the steps to solve $x^2 + 6x + 7 = 0$ by completing the square on sticky notes. Unfortunately he drops them and they get out of order. Put the notes into the correct order.

   - Add 2 to both sides
   - Subtract 3 from both sides
   - Write $x^2 + 6x + 7 = 0$ as $(x + 3)^2 - 2 = 0$
   - Take the square root of both sides

TARGETING DISTINCTION

4. Solve these equations by completing the square.
   a) $2x^2 - 6x - 3 = 0$
   b) $4x^2 - 8x + 1 = 0$
   c) $2x^2 + 5x - 10 = 0$
   d) $0.5x^2 - 7.5x + 8 = 0$

5. Solve these equations by completing the square.
   a) $-x^2 - 2x + 9 = 0$
   b) $-2x^2 - 6x + 3 = 0$
   c) $-3x^2 - 8x + 10 = 0$
   d) $-2x^2 - 5x + 6 = 0$

6. Show, by completing the square, why the equation $-2x^2 - 5x - 6 = 0$ has no real roots.

TARGETING DISTINCTION*

7. a) Write the equation $y = -2x^2 + 3x + 6$ in the form $a - b(x + c)^2$, where a, b and c are constants to be found.
   b) Solve the equation $-2x^2 + 3x + 6 = 0$, writing your answers in surd form
   c) Sketch the graph of $y = -2x^2 + 3x + 6$ labelling the minimum point and any points where the graph intersects the coordinate axes.

6.3 Solve simultaneous equations

THIS SECTION WILL SHOW YOU HOW TO ...
✓ solve two linear simultaneous equations in two variables
✓ solve two simultaneous equations in two variables one linear and one quadratic

KEY WORDS
✓ simultaneous equations   ✓ substitution method
✓ variable   ✓ non-linear equation

The balancing coefficients method for solving two linear simultaneous equations

Solve these **simultaneous equations**.
$$4x + 3y = 27 \quad (1)$$
$$5x - 2y = 5 \quad (2)$$

Both equations have to be changed to obtain identical terms in either x or y.

To make the y-coefficients the same, you can multiply the first equation by 2 (the y-coefficient of the second equation) and the second equation by 3 (the y-coefficient of the first equation) and then add the equations.

> **HINTS AND TIPS**
>
> It is also possible to make the x-coefficients the same, but you would need to subtract one equation from the other. This can lead to errors when dealing with negative coefficients, so it is safer to use the method that involves adding rather than subtracting.

Step 1: $(1) \times 2$ or $2 \times (4x + 3y = 27) \Rightarrow 8x + 6y = 54 \quad (3)$
$\qquad\quad (2) \times 3$ or $3 \times (5x - 2y = 5) \Rightarrow 15x - 6y = 15 \quad (4)$
$\qquad\quad$ Label the new equations (3) and (4).
Step 2: Eliminate one of the variables: $(3) + (4) \Rightarrow 23x = 69$
Step 3: Solve the equation: $\qquad\qquad\qquad\qquad\qquad x = 3$
Step 4: Substitute into equation (1): $\qquad\qquad\qquad 12 + 3y = 27$
Step 5: Solve the equation: $\qquad\qquad\qquad\qquad\qquad y = 5$
Step 6: Check: using (1) $4 \times 3 + 3 \times 5 = 12 + 15 = 27$ and using (2)
$\qquad\quad 5 \times 3 - 2 \times 5 = 15 - 10 = 5$

These are correct, so the solution is $x = 3$ and $y = 5$.

The substitution method for solving two linear simultaneous equations

There are five steps in the **substitution method**.
Step 1: Rearrange one of the equations into the form $y = \ldots$ or $x = \ldots$.
Step 2: Substitute the right-hand side of this equation into the other equation in place of the **variable** on the left-hand side.
Step 3: Expand and solve this equation.
Step 4: Substitute the value into the $y = \ldots$ or $x = \ldots$ equation.

Step 5: Check that the values work in both original equations.

The method you use depends very much on the coefficients of the variables and the way that the equations are written in the first place.

EXAMPLE 12

Solve these simultaneous equations.

$$y = 2x + 3$$
$$3x + 4y = 1$$

Because the first equation is in the form $y = \ldots$ the substitution method can easily be used.

Label the equations to help you follow through the method.

$$y = 2x + 3 \quad (1)$$
$$3x + 4y = 1 \quad (2)$$

Step 1: As equation (1) is in the form $y = \ldots$ there is no need to rearrange an equation.
Step 2: Substitute the right-hand side of equation (1) into equation (2) for the variable y.

$$3x + 4(2x + 3) = 1$$

Step 3: Expand and solve the equation.
$$3x + 8x + 12 = 1$$
$$11x = -11$$
$$x = -1$$

Step 4: Substitute $x = -1$ into $y = 2x + 3 \Rightarrow y = -2 + 3 = 1$
Step 5: Test the values in $y = 2x + 3$, which gives $1 = -2 + 3$, and in $3x + 4y = 1$, which gives $-3 + 4 = 1$.

These are correct so the solution is $x = -1$ and $y = 1$.

EXAMPLE 13

Solve these simultaneous equations by substitution.

$$3x + y = 5 \quad (1)$$
$$5x - 2y = 12 \quad (2)$$

Step 1: Rearrange (1) $\quad y = 5 - 3x \quad (3)$
Step 2: Substitute (3) into (2) $\quad 5x - 2(5 - 3x) = 12$
Step 3: Expand and solve $\quad 5x - 10 + 6x = 12$
$$11x = 22$$
$$x = 2$$
Step 4: Substitute back: $\quad 3 \times 2 + y = 5 \Rightarrow y = -1$
Step 5: Test: (1) $3 \times 2 - 1 = 5$ and (2) $5 \times 2 - 2 \times (-1) = 10 + 2 = 12$.

These are correct, so the solution is $x = 2$ and $y = -1$.

EXAMPLE 14

Solve these simultaneous equations. (One equation is linear and one is **non-linear**.)

$$x^2 + y^2 = 5$$
$$x + y = 3$$

When you have a linear and a non-linear equation you use the substitution method to solve them.

Call the equations (1) and (2):

$$x^2 + y^2 = 5 \quad (1)$$
$$x + y = 3 \quad (2)$$

> **HINTS AND TIPS**
>
> You should always give answers as a pair of values in x and y.

Rearrange the linear equation (2): $\quad x = 3 - y$

Substitute this into equation (1): $\quad (3 - y)^2 + y^2 = 5$

Expand and rearrange into the $\quad 9 - 6y + y^2 + y^2 = 5$

general form of a quadratic equation: $\quad 2y^2 - 6y + 4 = 0$

Divide by 2: $\quad y^2 - 3y + 2 = 0$

Factorise: $\quad (y - 1)(y - 2) = 0$

Solve: $\quad y = 1 \text{ or } y = 2$

To find x, substitute both values for y into equation (2).
When $y = 1$, $x = 3 - 1 = 2$ and when $y = 2$, $x = 3 - 2 = 1$.
These answers can be written as the coordinates (2, 1) and (1, 2).

EXERCISE 6E

TARGETING PASS

1. Use the substitution method to solve each pair of linear simultaneous equations.

   a) $2x + y = 9$
      $x - 2y = 7$

   b) $3x - 2y = 10$
      $4x + y = 17$

   c) $x - 2y = 10$
      $2x + 3y = 13$

TARGETING MERIT

2. Solve each pair of simultaneous equations.

   a) $xy = 2$
      $y = x + 1$

   b) $xy = -4$
      $2y = x + 6$

3. Solve each pair of simultaneous equations.

   a) $x^2 + y^2 = 25$
      $x + y = 7$

   b) $x^2 + y^2 = 9$
      $y = x + 3$

   c) $x^2 + y^2 = 13$
      $5y + x = 13$

4. Solve each pair of simultaneous equations.

   a) $y = x^2 + 2x - 3$
      $y = 2x + 1$

   b) $y = x^2 - 2x - 5$
      $y = x - 1$

   c) $y = x^2 - 2x$
      $y = 2x - 3$

5. Solve these pairs of simultaneous equations.

a) $y = x^2 + 3x - 3$ and $y = x$

b) $x^2 + y^2 = 13$ and $x + y = 1$

c) $x^2 + y^2 = 5$ and $y = x + 1$

d) $y = x^2 - 3x + 1$ and $y = 2x - 5$

EXAMPLE 15

a) Solve these simultaneous equations giving your answer in surd form.

$x^2 + y^2 = 12$

$x + y = 1$

b) Sketch the graphs of the two equations on the same axes labelling the coordinates of any points of intersection.

a) Call the equations (1) and (2):

$x^2 + y^2 = 12$ (1)

$x + y = 1$ (2)

Rearrange the linear equation (2): $x = 1 - y$

Substitute this into equation (1): $(1 - y)^2 + y^2 = 12$

Expand and rearrange into the $1 - 2y + y^2 + y^2 = 12$

general form of the quadratic equation: $2y^2 - 2y - 11 = 0$

Solve by using the quadratic formula: $\dfrac{-b \pm \sqrt{b^2 - 4ac}}{2a}$

$a = 2$, $b = -2$ and $c = -11$

$$y = \frac{2 \pm \sqrt{(-2)^2 - 4(2)(-11)}}{2 \times 2}$$

$$= \frac{2 \pm \sqrt{4 + 88}}{4}$$

$$= \frac{2 \pm \sqrt{92}}{4}$$

$$= \frac{1 \pm \sqrt{23}}{2}$$

To find x, substitute both values of y into equation (2).

When $y = \dfrac{1 + \sqrt{23}}{2}$, $x = \dfrac{1 - \sqrt{23}}{2}$ and when $y = \dfrac{1 - \sqrt{23}}{2}$, $x = \dfrac{1 + \sqrt{23}}{2}$

These answers can be written as the coordinates $\left(\dfrac{1 - \sqrt{23}}{2}, \dfrac{1 + \sqrt{23}}{2}\right)$ and $\left(\dfrac{1 + \sqrt{23}}{2}, \dfrac{1 - \sqrt{23}}{2}\right)$.

b) The graph of $x^2 + y^2 = 12$ is a circle with centre (0, 0) and radius $\sqrt{12}$.

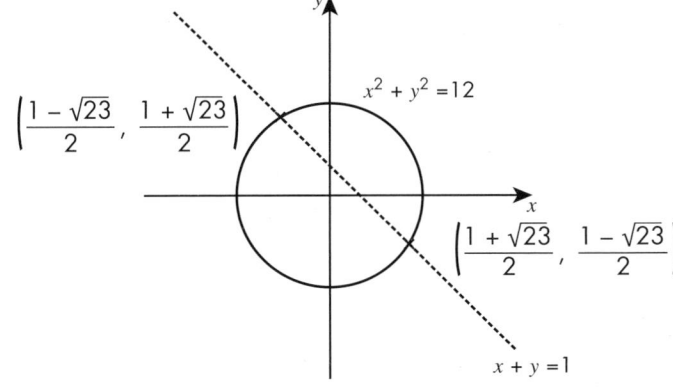

HINTS AND TIPS

Remember that a sketch does not have to have exact points plotted. You just need to label the axes, the two graphs and the points of intersection.

TARGETING DISTINCTION

6. $x^2 = \frac{4}{3}y$ and $x - y = -1$

   a) Solve this pair of simultaneous equations

   b) Sketch the graphs of the two equations on the same axes. Label the exact coordinates of any points of intersection.

7. $x^2 + y^2 = 8$ and $y - 2x - 2 = 0$

   a) Solve this pair of simultaneous equations

   b) Sketch the graphs of the two equations on the same axes. Label the exact coordinates of any points of intersection.

TARGETING DISTINCTION*

8. Solve these pairs of simultaneous equations. Write you answers in surd form, where appropriate.

   a) $-x^2 + 3y^2 - 2x = 11$ and $x = -2y - 3$

   b) $x^2 + y^2 - 2x = 20$ and $y - x = -1$

Problem solving using simultaneous equations

EXAMPLE 16

Two numbers have a sum of $\frac{21}{2}$ and a product of 20.

By forming and solving two simultaneous equations find the two numbers.

Form two equations to represent the problem.

Call the equations (1) and (2): $x + y = \frac{21}{2}$ (1)

$xy = 20$ (2)

Rearrange the linear equation (1) to make x the subject: $x = \frac{21}{2} - y$

Substitute this into equation (2): $\left(\frac{21}{2} - y\right)y = 20$

Expand and rearrange into the $\frac{21}{2}y - y^2 = 20$

general form of the quadratic equation: $y^2 - \frac{21}{2}y + 20 = 0$

Multiply by 2: $2y^2 - 21y + 40 = 0$

Factorise: $(2y - 5)(y - 8) = 0$

Solve: $y = \frac{5}{2}$ or $y = 8$

Substitute both values for y in equation (2).

When $y = = \frac{5}{2}$, $x = 8$ and when $y = 8$, $x = \frac{5}{2}$.

So, the two numbers are $= \frac{5}{2}$ and 8.

EXERCISE 6F

TARGETING PASS

1. By forming and solving two simultaneous equations find the two numbers for each question.

  a) The sum of two numbers equals 4 and the product equals $\frac{7}{4}$.

  b) The sum of two numbers equals 7 and the product of their squares equals 25.

  c) The difference of two numbers equals 5 and the difference of their squares equals 45.

TARGETING MERIT

2. Do the line $x + y = \frac{5}{3}$ and the curve $x^2 + 2xy = 6$ intersect?

Explain your answer.

TARGETING DISTINCTION

3. The diagram shows a trapezium and a triangle (not drawn to scale).

The area of the trapezium is 21 cm², the area of the triangle is 28 cm²

By forming and solving two simultaneous equations, find the value of x and the value of y where $y \in \mathbb{Z}$.

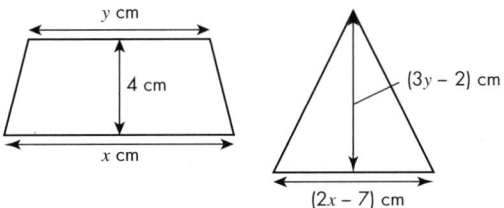

TARGETING DISTINCTION*

4. The diagram shows a circle and a line.

  a) Find the equation of the (i) circle and (ii) the line.

  b) Solve these equations simultaneously to find the coordinates of A and B.

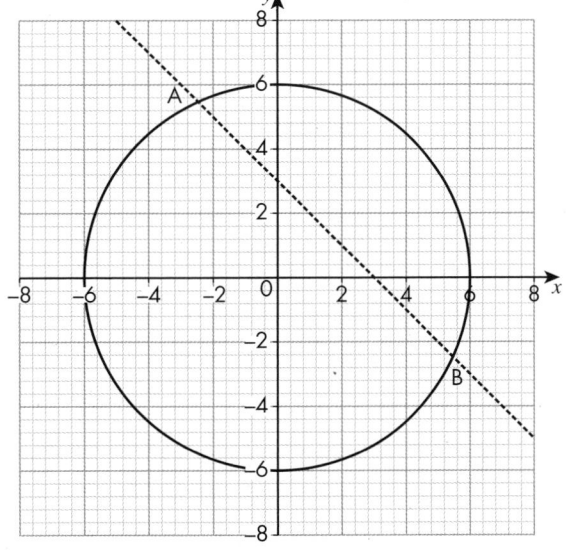

6.4 Solving inequalities

THIS SECTION WILL SHOW YOU HOW TO ...
- solve linear inequalities in one or two variables
- solve quadratic inequalities in one variable
- represent solutions sets on number lines using set notation, and on a graph
- answer inequality questions in context such as in linear programming and profit and loss problems

KEY WORDS
- inequality
- solution set
- linear programming
- quadratic inequality
- number line
- feasible region
- objective function
- vertex (plural, vertices)

Inequalities behave similarly to equations. You use the same rules to solve linear inequalities as you use for linear equations.

EXAMPLE 17

Solve $\frac{x}{2} + 4 \geqslant 13$.

Solve just like an equation, but leave the inequality sign in place of the equals sign.
Subtract 4 from both sides: $\quad \frac{x}{2} \geqslant 9$
Multiply both sides by 2: $\quad x \geqslant 18$
This means that x can take any value above and including 18.

If you multiply or divide by a negative number when you are solving an inequality, you must change the inequality sign. This means that:

- 'less than' becomes 'more than'
- 'more than' becomes 'less than'.

EXAMPLE 18

Solve the inequality $10 - 2x \geqslant 3$.

Subtract 10 from both sides: $\quad -2x \geqslant -7$
Divide both sides by -2: $\quad x \leqslant 3.5$ Note that the inequality sign has been reversed.
You could approach this example in a different way.
$$10 - 2x \geqslant 3$$
Add $2x$ to both sides: $\quad 10 \geqslant 2x + 3$
Subtract 3 from both sides: $\quad 7 \geqslant 2x$
Divide both sides by 2: $\quad 3.5 \geqslant x$
The sign does not change this time. $3.5 \geqslant x$ is equivalent to $x \leqslant 3.5$.

> **HINTS AND TIPS**
>
> You must reverse inequality signs when multiplying or dividing both sides by a negative number.

EXAMPLE 19

Solve the inequality $-3 < 2x + 3 < 14$.

For compound inequalities solve each side separately:

$$-3 < 2x + 3 \qquad 2x + 3 < 14$$
$$-6 < 2x \qquad 2x < 11$$
$$-3 < x \qquad x < \frac{11}{2}$$

Finally bring the two inequalities together.

This gives an answer of $-3 < x < \frac{11}{2}$.

EXERCISE 6G

TARGETING PASS

1. Solve these linear inequalities.
    a) $4x + 1 \geqslant 3x - 5$
    b) $3y - 12 \leqslant y - 4$
    c) $2(4x - 1) \leqslant 3(x + 4)$

2. Solve these inequalities.
    a) $20 - 2x \leqslant 5$
    b) $3 - 4x \geqslant 11$
    c) $25 - 3x > 7$
    d) $2(6 - x) < 9$
    e) $\frac{10 - 2x}{5} \leqslant 4$
    f) $\frac{8 - 4x}{3} > 2$

3. Solve these linear inequalities.
    a) $7 < 2x + 1 < 13$
    b) $5 < 3x - 1 < 14$
    c) $-1 \leqslant 5x + 4 \leqslant 19$
    d) $1 \leqslant 4x - 3 < 13$
    e) $11 \leqslant 3x + 5 < 17$
    f) $-3 \leqslant 2x - 3 \leqslant 7$

Suppose that $x^2 > 16$. What can you say about x? This is a **quadratic inequality**.

x has two solutions which you can see by factorising $x^2 - 16 = 0$.

This is the difference of two squares, so it factorises to $(x + 4)(x - 4) = 0$,

So the solutions are $x = 4$ and $x = -4$.

You can represent these solutions on a **number line**.

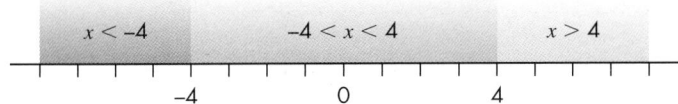

By choosing a value in each section in turn, and squaring it, you can see that:

if $x > 4$, then $x^2 > 16$ (e.g. $5^2 = 25 > 16$)

if $-4 < x < 4$, then $x^2 < 16$ (e.g. $2^2 = 4 < 16$)

if $x < -4$, then $x^2 > 16$ (e.g. $(-5)^2 = 25 > 16$)

So the **solution set** for $x^2 > 16$ is in two parts: $x < -4$ and $x > 4$.

This can be shown on a number line.

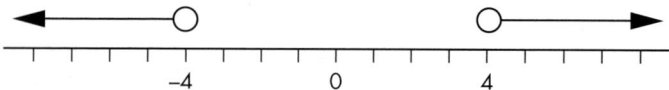

Notice that the boundary values (−4 and 4) *are not* included because $4^2 = (-4)^2 = 16$, so they are shown with open circles.

Neither of them is less than 16.

The solution to $x^2 \geqslant 16$ is similar but, in this case, the boundary values *are* included because the inequality includes 16 as a possible value.

So the solution set for $x^2 \geqslant 16$ is in two parts: $x \leqslant -4$ and $x \geqslant 4$.

The boundary values (−4 and 4) are shown with filled circles.

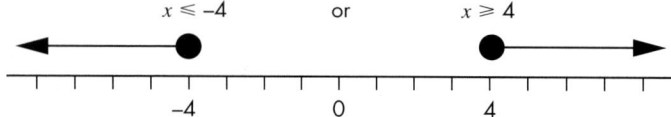

These solutions can also be represented graphically by sketching $x^2 - 16 = 0$.

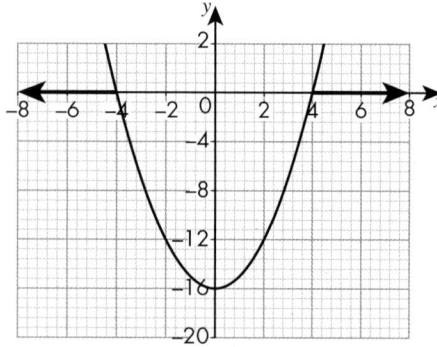

For $x^2 - 16 > 0$, the solution is $\{x: x < -4\} \cup \{x: x > 4\}$.

(The symbol $\cup$ is used here to denote a 'union' between the two defined solutions.)

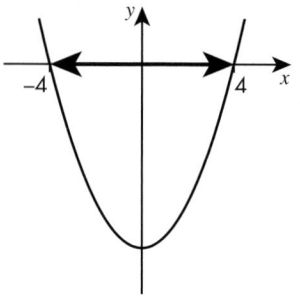

For $x^2 - 16 < 0$, the solution is continuous: $\{x: -4 < x < 4\}$.

EXAMPLE 20

Solve the inequality $x^2 - x - 6 < 0$.

Factorising the quadratic gives $(x + 2)(x - 3) < 0$.

Sketching the graph and identifying the solution set gives:

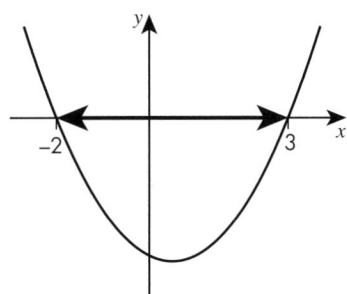

So, for $(x + 2)(x - 3) < 0$, the solution is continuous: $\{x: -2 < x < 3\}$.

EXERCISE 6H

TARGETING PASS

1. Solve these inequalities.
   a) $x^2 \leq 16$ b) $x^2 < 4$ c) $x^2 > 6.25$ d) $x^2 \geq 1$

2. Solve these inequalities.
   a) $2x^2 < 18$ b) $3x^2 > 75$ c) $4x^2 \geq 9$ d) $4x^2 \leq 1$

3. Solve these inequalities.
   a) $x^2 - 4 \leq 0$ b) $x^2 - 12.25 > 0$
   c) $8x^2 - 50 < 0$ d) $9 - x^2 \geq 0$

HINTS AND TIPS

Remember to factorise first and then sketch a graph to help identify the solutions.
Always use the same inequality in the question for your answer.

TARGETING MERIT

4. Solve each inequality.
   a) $x^2 + 2x - 3 \leq 0$ b) $3x^2 + 7x > 6$ c) $9x^2 + 4x < 5$
   d) $5x^2 + 22x \leq 15$ e) $3x^2 \leq 11x + 4$

5. Work out the integer values of x that satisfy the inequality $3x^2 - 19x + 20 < 0$.

6. Work out the integer values of x that satisfy the inequality $4x^2 - 12x + 5 < 0$.

7. Solve these inequalities.

a) $x^2 - 3x - 10 > 0$
b) $x^2 + 12x + 35 < 0$
c) $x^2 \leq x + 72$
d) $3x^2 - 10x + 3 \leq 0$
e) $2x^2 + 13x + 11 > 0$
f) $5x^2 + 6 \geq 13x$

Graphical inequalities

Inequalities can be represented on graphs to solve problems.

EXAMPLE 21

Show each of these inequalities on a graph, shading each of the required regions.
a) i) $x \geq 3$ ii) $y < -1$ iii) $2x + y > 6$ iv) $3x - 2y \leq 0$
b) List all the points in the unshaded region where x and y are both integers.

    i) Draw the line $x = 3$. It is solid because points on the line are included in the required region.

    ii) Draw the line $y = -1$. It is dashed because points on the line are not included in the required region.

    iii) Draw the line $2x + y = 6$. It is dashed because points on the line are not included in the required region.

    iv) Draw the line $3x - 2y = 0$. It is solid because points on the line are included in the required region.

Check each region using (0, 0) before shading

    i) $0 < 3$, which is not ≥ 3, so the origin is not in the required region.

    ii) $0 > -1$, so the origin is not in the required region.

    iii) $2(0) + 0 < 6$, so $2x + y$ is not > 6 and the origin is not in the required region.

    iv) $3(0) - 0 = 0$, so the origin will lie on boundary of the region.

Also checking (0, 1) for (iv) gives $3(0) - 2(1) < 0$ and so (0, 1) will lie in the required region.

a)
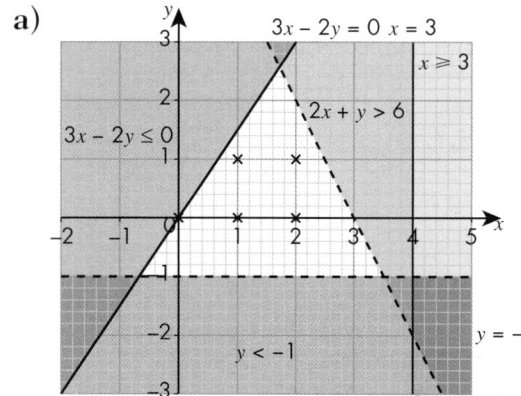

b) The crosses mark all the points in the unshaded region where x and y are both integers.
These are (0, 0), (1, 0), (2, 0), (1, 1) and (2, 2).

Points that are on broken/dashed lines are not included, whereas points on solid lines are included.

EXERCISE 6I

TARGETING PASS

1. Show each of these inequalities on a graph shading each of the defined regions.

   a) i) $x < -3$ ii) $y \geqslant 4.5$ iii) $x + y > 4$ iv) $y - x \leqslant 3.5$

   b) List all the points in the unshaded region where x and y are both integers.

2. Write down the inequalities that define the unshaded region.

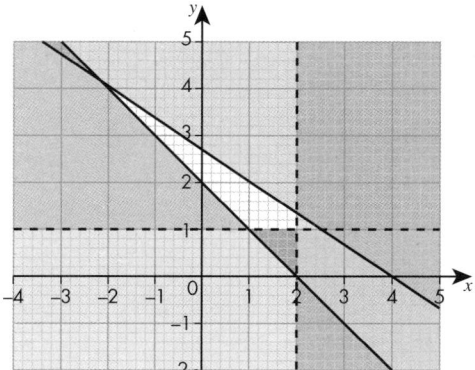

HINTS AND TIPS

Remember to check your shading by testing (0, 0) to find out whether or not it is in the required region.
For question 2 you are defining the unshaded region so your inequalities need to show this.

Linear programming

Linear programming is used to help solve real-life problems.

Inequalities are used to represent constraints to find optimal solutions. The region that satisfies all the inequalities is called the **feasible region**. If the unwanted regions are shaded, then the feasible region is the *unshaded* region.

The **objective function** describes the quantity required to be either a maximum (e.g. profit) or minimum (e.g. costs) in a linear programming problem.

EXAMPLE 22

At a theme park, the big wheel ride costs £5 and a ride on the roller coaster £7.50. Amina has £25 to spend. She has X big wheel rides and Y rides on the roller coaster

a) Use this information to form five inequalities.

b) Graph these inequalities using the horizontal axis for X and the vertical axis for Y

c) Mark all the possible combinations of rides that satisfy these conditions

d) If Amina wants to go on as many rides as possible and go on both rides but prefers the big wheel, what combination should she choose and what would this cost her?

   a) $X \geqslant 0$ (as the number of rides cannot be negative)

   As Amina has £25 $5X \leqslant 25 \Rightarrow X \leqslant 5$ so $0 \leqslant X \leqslant 5$

   $Y \geqslant 0$ (as the number of rides cannot be negative)

   As Amina has £25 $7.5Y \leqslant 25 \Rightarrow Y \leqslant \frac{10}{3}$ so $0 \leqslant Y \leqslant \frac{10}{3}$

   The total cost of the rides cannot exceed £25,

   so $5X + 7.5Y \leqslant 25$ $\Rightarrow 2X + 3Y \leqslant 10$

b) The line $X = 0$ is the y-axis. Shade the unwanted region $X \leq 0$.

Draw the solid line $X = 5$ and shade $X \geq 5$.

The line $Y = 0$ is the x-axis. Shade the unwanted region $Y \leq 0$.

Draw the solid line $Y = \frac{10}{3}$ and shade $Y \geq \frac{10}{3}$.

Draw the solid line $2X + 3Y = 10$ and shade $2X + 3Y \geq 10$.

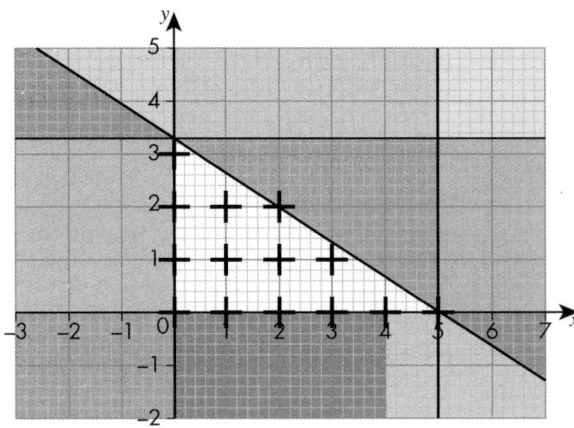

c) the possible combinations are marked on the graph

d) Amina should go on three big wheel rides and one roller coaster ride.

This would cost her $3 \times 5 + 1 \times 7.5 = 15 + 7.5 = £22.50$

TARGETING DISTINCTION

3. Mr Foster is buying pens and pencils for his students and has £15 to spend.

The pens cost 50p each and the pencils cost 25p each.

He must buy more than 40 pens and pencils altogether and must buy at least 10 pens and at least 20 pencils.

a) Let the number of pens he buys be x and the number of pencils y. Use this information to form four inequalities

b) Graph these inequalities using the horizontal axis for x and the vertical axis for y

c) Indicate on your graph the region containing all the possible combinations of pens and pencils that satisfy these conditions

d) If Mr Foster wants the total number of pens and pencils to be as large as possible, what combination of pens and pencils should he choose?

The maximum or minimum values of the objective function will lie at, or near to, one of the **vertices** of the feasible region.

EXAMPLE 23

Lily is making bags and cushions to sell at a market.

She makes some bags (x) and some cushions (y).

Each bag costs £3 to make and each cushion costs £5 to make.

Lily had a budget of £500 to spend on making the bags and cushions.

a) Write down an inequality, in terms of x and y, to model this.

Two further constraints (limitations) are:

$y \leq 2x$ and $5y - x \geq 100$

b) Draw a graph to represent all of the inequalities and information given.

Lily's objective is to make as many bags and cushions (in total) as possible.

c) State the objective function.

d) Using your graph determine the optimal number of bags and cushions that Lily should make.

a) $3x + 5y \leq 500$

b)

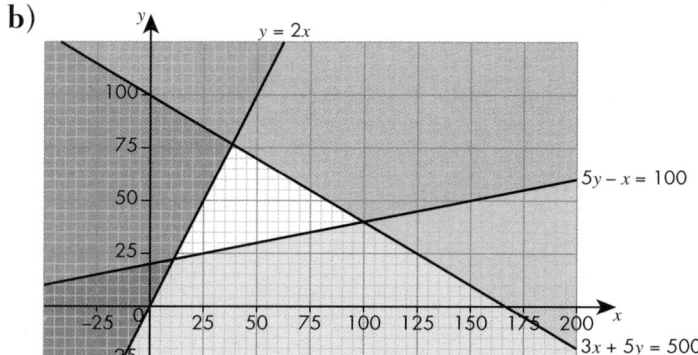

Lily's objective is to make as many bags and cushions (in total) as possible.

c) The objective is to maximise $(P) = x + y$

d) The optimal number of bags and cushions can be found at a vertex (where two lines meet) in the feasible (unshaded) region.

By testing each one of the three the maximum value of $(P) = x + y$ is found at the point (100, 40) which represents 100 bags and 40 cushions, a total of 140 items altogether.

TARGETING DISTINCTION*

4. In a sports competition there are two types of prizes, medals and trophies.

The organisers have decided that they will award at least 25 medals (x) and fewer than 60 trophies (y).

a) Write down the two inequalities to model these constraints.

Two further constraints are

$3x - 2y \leq 90$ and $2x + 5y > 250$

b) Draw a graph to represent all four of these inequalities and label the feasible region R.

The cost of a trophy is double the cost of a medal and the organisers want to minimise the cost of the prizes.

c) State the objective function in terms of x and y.

d) Use the graph to find the number of medals and trophies the organisers should award.

EXAMPLE 24

On a graph show the region that satisfies the inequalities

$$x^2 - 3x - 4 > y \quad \text{and} \quad y < x + 5$$

Factorise $x^2 - 3x - 4 = (x + 1)(x - 4)$ to find where the graph cuts the x-axis.

The x-intercepts are $(-1, 0)$ and $(4, 0)$.

The y-intercept is $(0, -4)$ as $y = -4$ when $x = 0$.

The minimum point of the quadratic graph can be found by completing the square:

$$x^2 - 3x - 4 = \left(x - \tfrac{3}{2}\right)^2 - \tfrac{9}{4} - 4$$
$$= \left(x - \tfrac{3}{2}\right)^2 - \tfrac{25}{4}$$

The minimum point of the quadratic graph is $\left(\tfrac{3}{2}, -\tfrac{25}{4}\right)$

Note that both lines are broken as the inequalities do not include the values on the lines.

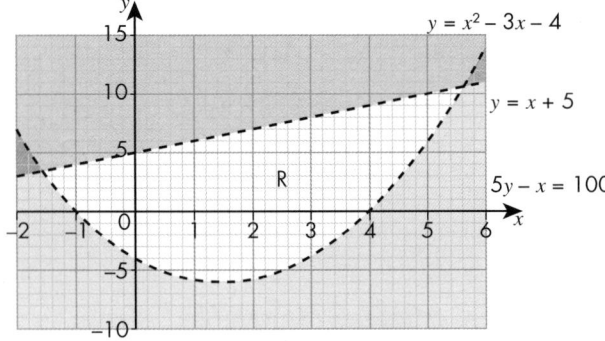

TARGETING DISTINCTION*

5. On a graph show the region that satisfies the inequalities

   $-2x^2 - 3x + 14 > y \quad \text{and} \quad y > x + 4$

Exam-style questions

1. Solve the equation $7\sqrt{x} = 30 - x$. [3 marks]

2. $\dfrac{5}{\sqrt{x}} \div \dfrac{3}{x} = \dfrac{5}{2}$ [3 marks]

3. Solve the equation $4 \sin x + 1 = 3$ for $0° \leqslant x \leqslant 360°$. [2 marks]

4. The length of a room is 2 m more than its width. The area of the room is 11 m².

   Work out the dimensions of the room giving your answer to 2 decimal places. [3 marks]

5. Find the coordinates of intersection between the circle
   $x^2 + y^2 - 6x + 10y + 9 = 0$ and the line $x + y = 3$. [3 marks]

6. a) Find the values of a, b and c such that
   $3x^2 - 12x + 4 \equiv a(x+b)^2 + c$. [3 marks]

   b) Hence, or otherwise, solve the equation $3x^2 - 12x + 4 = 0$, giving your solutions correct to 2 decimal places.

7. ABCD is a rectangle.
   a) The area of the rectangle is at least 160 cm².
   Use this fact to set up an inequality for x. [1 mark]

   b) Solve the inequality and hence find the smallest possible length of AD. [4 marks]

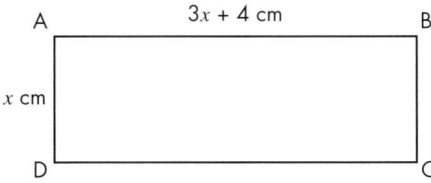

Not to scale [3 marks]

8. Solve the inequality $(x+2)^2 < (x-6)^2$. [4 marks]

9. Find the integer solutions of the inequality $N^2 - 2N < 8$. [4 marks]

10. Solve the inequality $12x < x^2 + 20$. [3 marks]

11. a) On the same set of axes, sketch the graphs of $y = x^2 - 3$ and $y = 2x$. [3 marks]

    b) Use the graphs to solve the inequality $x^2 - 3 \leq 2x$. [1 mark]

7 Pythagoras' theorem and trigonometry

7.1 Pythagoras' theorem in 2D and 3D

THIS SECTION WILL SHOW YOU HOW TO ...

✓ identify appropriate right-angled triangles in 2- and 3-dimensional shapes and apply Pythagoras' theorem

KEY WORDS

✓ hypotenuse ✓ isosceles triangle ✓ Pythagoras' theorem ✓ 3D

Pythagoras' theorem can be written as a formula: $c^2 = a^2 + b^2$

Remember that Pythagoras' theorem can only be used in right-angled triangles.

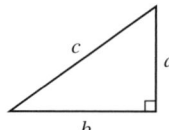

EXAMPLE 1

Calculate the area of this triangle. Give your answer to 1 decimal place.

To find the area, you first need to calculate the height of the triangle.
The triangle is isosceles so you can split it into two right-angled triangles.

Let the height be x cm.

Then, using Pythagoras' theorem:
$$\begin{aligned} x^2 &= 7.5^2 - 3^2 \\ &= 56.25 - 9 \\ &= 47.25 \\ x &= \sqrt{47.25} = 6.8738 \ldots \end{aligned}$$

Keep all the figures on the calculator.

The area of the triangle is $\frac{1}{2} \times 6 \times 6.87\,\text{cm}^2 = 20.6\,\text{cm}^2$ (1 decimal place).

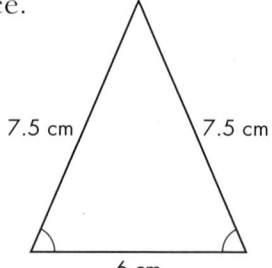

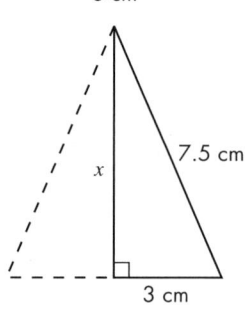

EXERCISE 7A

TARGETING PASS

1. Calculate the area of each of these isosceles triangles.

a)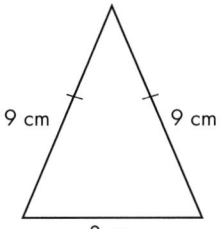
b)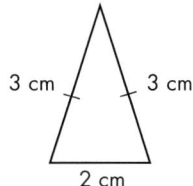
c)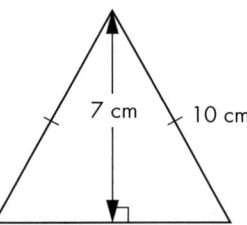

Pythagoras' theorem in three dimensions

You can use Pythagoras' theorem to solve problems in **3D** in exactly the same way as for 2D problems.

- Identify the right-angled triangle you need.
- Redraw this triangle and label it with the given lengths and the length to be found, usually x or y.
- From your diagram, decide whether you need to find the **hypotenuse** or one of the shorter sides.
- Solve the problem, rounding to a suitable degree of accuracy.

EXAMPLE 2

What is the longest piece of straight wire that can be stored in this box measuring 30 cm by 15 cm by 20 cm? Give your answer to 1 decimal place.

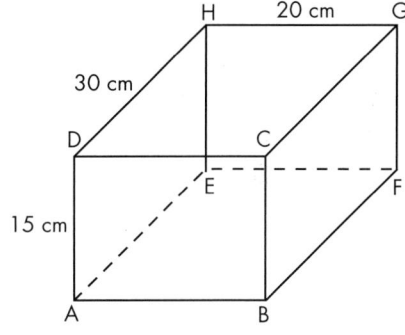

The longest distance across this box is any one of the diagonals AG, DF, CE or BH.

In this case, take AG.

First, identify a right-angled triangle containing AG and draw it.

This gives a triangle AFG, which contains two lengths you do not know, AG and AF.

Let AG = x cm and AF = y cm.

Next identify a right-angled triangle that contains the side AF and draw it.

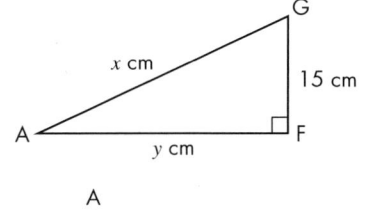

This gives a triangle ABF. You can now find AF.

By Pythagoras' theorem: $y^2 = 30^2 + 20^2 = 1300$ (there is no need to find y.)

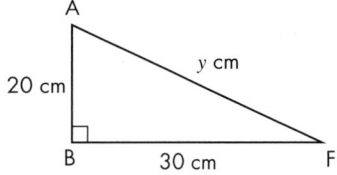

Now use triangle AFG to find AG.

By Pythagoras' theorem: $x^2 = y^2 + 15^2$
$= 1300 + 225$
$= 1525$

So $x = \sqrt{1525} = 39.1$ (1 dp).

Therefore, the longest straight wire that can be stored in the box is 39.1 cm.

TARGETING PASS

2. A box measures 8 cm by 12 cm by 5 cm.
   a) Calculate the length of: **i)** AC   **ii)** BG   **iii)** BE.
   b) Calculate the diagonal distance BH.

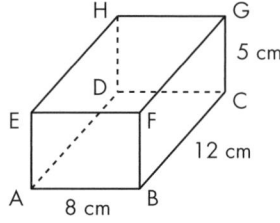

3. A garage is 5 m long, 3 m wide and 3 m high. Can a 7 m long pole be stored in it? Show how you worked out your answer.

4. Spike, a spider, is at the corner S of the wedge shown in the diagram. Fred, a fly, is at the corner F of the same wedge.

   a) Calculate the shortest distance Spike would have to travel to get to Fred, if she used the edges of the wedge.
   b) Calculate the distance Spike would have to travel across the face of the wedge to get directly to Fred.

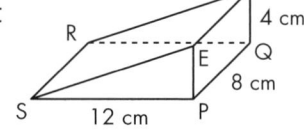

5. Fred is now at the top of a baked beans can and Spike is directly below him on the base of the can. To catch Fred by surprise, Spike takes a diagonal route round the can.

   How far does Spike travel?

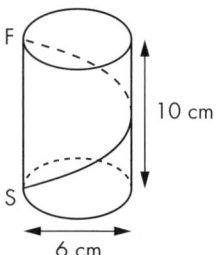

6. If each side of a cube is 10 cm long, how far will it be from one corner of the cube to the opposite one?

7. A pyramid has a square base of side 20 cm and each sloping edge is 25 cm long.

   Find the height of the pyramid.

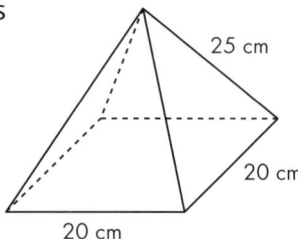

8. The diagram shows a cuboid with sides of 40 cm, 30 cm and 22.5 cm.

   M is the midpoint of the side FG. Calculate (or write down) these lengths, giving your answers to 3 significant figures, if necessary.

   a) AH   **b)** AG   **c)** AM   **d)** HM

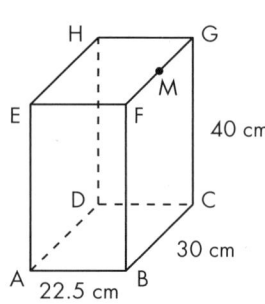

7.2 Trigonometry in 2D and 3D

THIS SECTION WILL SHOW YOU HOW TO ...
✓ identify appropriate right-angled triangles in 2- and 3-dimensional shapes and apply trigonometry
✓ work out the angle between a line and a plane
✓ work out the angle between two planes

KEY WORDS
✓ ratios   ✓ sine   ✓ cosine
✓ tangent   ✓ angle of elevation   ✓ angle of depression

In a right-angled triangle, the **sine**, **cosine** and **tangent ratios** for angle θ are defined as:

$$\text{sine } \theta = \frac{\text{opposite}}{\text{hypotenuse}} \quad \text{cosine } \theta = \frac{\text{adjacent}}{\text{hypotenuse}} \quad \text{tangent } \theta = \frac{\text{opposite}}{\text{adjacent}}$$

Adjacent

These ratios are usually abbreviated to:

$$\sin \theta = \frac{O}{H} \quad \cos \theta = \frac{A}{H} \quad \tan \theta = \frac{O}{A}$$

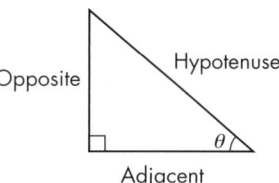

EXAMPLE 3

Work out the length AD giving your answer to one decimal place.
First find the length of the hypotenuse AC.

$$\cos \theta = \frac{\text{adjacent}}{\text{hypotenuse}} \Rightarrow \cos 43° = \frac{14}{AC} \Rightarrow AC = \frac{14}{\cos 43°}$$

= 19.14... (keep this value in your calculator for accuracy)

Next use this value to find AD.

$$\tan \theta = \frac{\text{opposite}}{\text{adjacent}} \Rightarrow \tan 51° = \frac{19.14...}{AD} \Rightarrow AD = \frac{19.14...}{\tan 51°}$$

$$= 15.5 \text{ cm (1 dp.)}$$

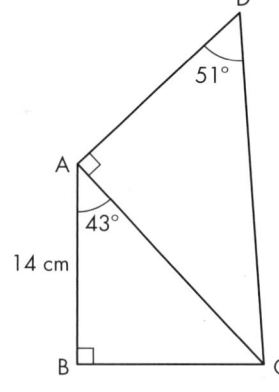

EXERCISE 7B

TARGETING PASS

In these questions, give answers involving angles to the nearest degree.

1. A ladder, 6 m long, rests against a wall. The foot of the ladder is 2.5 m from the base of the wall. What angle does the ladder make with the ground?

2. The ladder in question **1** has a 'safe angle' with the ground of between 70° and 80°. What are the safe limits for the distance of the foot of this ladder from the wall? How high up the wall does the ladder reach?

3. Calculate the angle that the diagonal makes with the long side of a rectangle that measures 10 cm by 6 cm.

4. Appleby is 800 km from Bixton on a bearing of 065°.

   a) How far north of Bixton is Appleby?
   b) How far west of Appleby is Bixton?

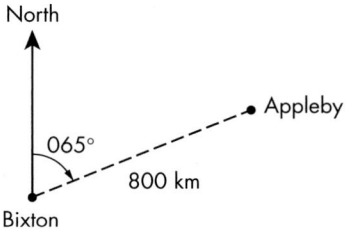

5. A ship is at S where it is 65 km from port P on a bearing of 132°.

   It sails north to port T where it is east of the port.

   Find the shortest distance from S to T.

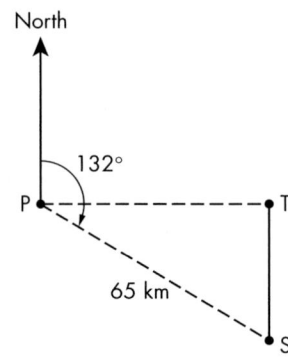

Angles of elevation and depression

When you look *up* at an aircraft in the sky, the angle through which your line of sight turns from looking straight ahead (the horizontal) is called the **angle of elevation**.

When you are standing on a high point and look *down* at a boat, the angle through which your line of sight turns from looking straight ahead (the horizontal) is called the **angle of depression**.

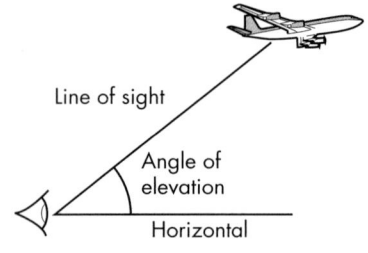
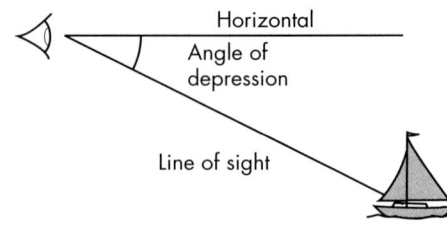

EXAMPLE 4

From the top of a vertical cliff, 100 m high, Ali sees a boat out at sea. The angle of depression from Ali to the boat is 42°. How far from the base of the cliff is the boat? Give your answer to 3 significant figures.

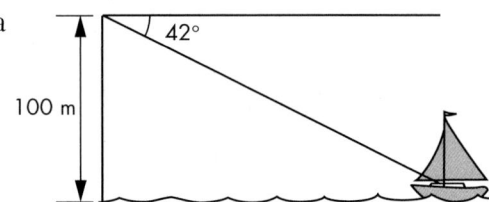

Draw a right-angled triangle and label what you know. You know the opposite side to the angle and want to find the adjacent side, so you need to use the tangent of the angle.

$\tan 42° = \dfrac{100}{x} \Rightarrow x = \dfrac{100}{\tan 42°}$
$= 111 \text{ m (3 sf)}$

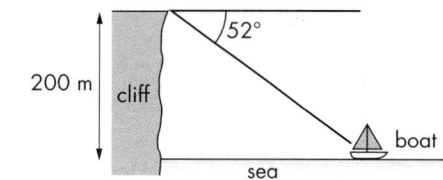

EXERCISE 7C

TARGETING PASS

In these questions, give any answers involving angles to the nearest degree.

1. Eric sees an aircraft in the sky. The aircraft is at a horizontal distance of 25 km from Eric. The angle of elevation is 22°. How high is the aircraft?

2. An aircraft is flying at an altitude of 4000 m and is 10 km from the airport. If a passenger can see the airport, what is the angle of depression?

3. a) From the top of a vertical cliff, 200 m high, a boat has an angle of depression of 52°. How far from the base of the cliff is the boat?

   b) The boat now sails away from the cliff so that the distance is doubled. Does that mean that the angle of depression is halved?

   Give a reason for your answer.

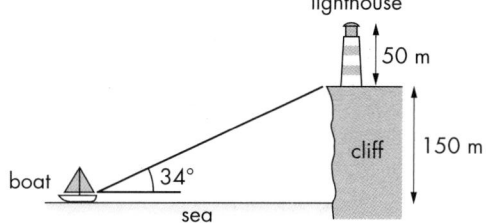

4. From a boat, the angle of elevation of the foot of a lighthouse on the edge of a cliff is 34°.

   a) If the cliff is 150 m high, how far from the base of the cliff is the boat?

   b) If the lighthouse is 50 m high, what would be the angle of elevation of the top of the lighthouse from the boat?

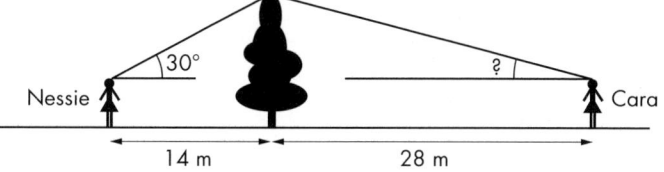

5. Nessie and Cara are standing on opposite sides of a tree.

   Nessie is 14 m away and the angle of elevation of the top of the tree is 30°.

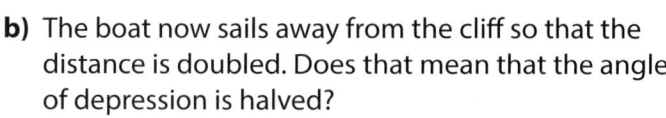

   Cara is 28 m away. She says the angle of elevation for her must be 15° because she is twice as far away.

   Is she correct?

   What do you think the angle of elevation is?

Problems in 3D

To find the value of an angle or side in a 3D shape, you need to find a right-angled triangle in the figure that contains it. This triangle also has to include two known values that you can use in the calculation.

You need to draw this triangle separately as a 2D, right-angled triangle. Label the known values and the unknown value that you want to find. Then use the trigonometric ratios and Pythagoras' theorem to solve the problem.

EXAMPLE 5

A, B and C are three points at ground level. They are in the same horizontal plane. C is 50 km east of B. B is north of A. C is on a bearing of 050° from A.

An aircraft, flying east, passes over B and over C at the same height. When it passes over B, the angle of elevation from A is 12°. Find the angle of elevation of the aircraft from A when it is over C. Give your answer to 1 decimal place.

First, draw a diagram and label all the known information.

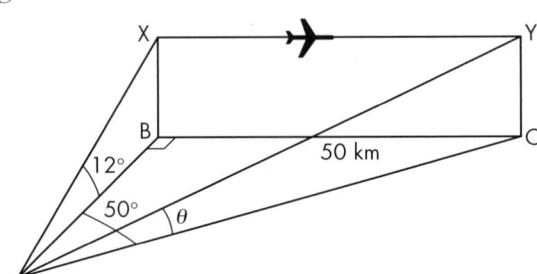

Next, use the right-angled triangle ABC to calculate AB and AC.

$$AB = \frac{50}{\tan 50°} = 41.95 \text{ km (4 significant figures)}$$

$$AC = \frac{50}{\sin 50°} = 65.27 \text{ km (4 significant figures)}$$

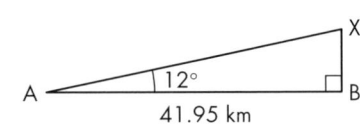

Then use right-angled triangle ABX to calculate BX, and hence CY.

BX = 41.95 tan 12° = 8.917 km (4 significant figures)

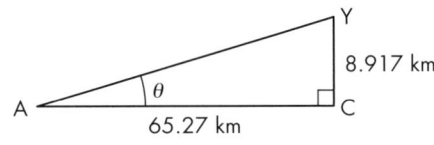

Finally, use the right-angled triangle ACY to calculate the required angle of elevation, θ.

$$\tan \theta = \frac{8.917}{65.27} = 0.1366$$

$$\Rightarrow \theta = \tan^{-1} 0.1366 = 7.8° \text{ (1 decimal place)}$$

EXERCISE 7D

TARGETING MERIT

1. A vertical flagpole AP stands at the corner of a rectangular courtyard ABCD.

   Calculate the angle of elevation of P from C.

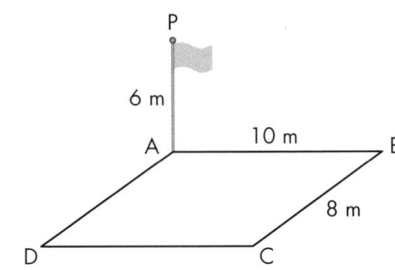

2. The diagram shows a pyramid. The base is a horizontal rectangle ABCD, 20 cm by 15 cm. The length of each sloping edge is 24 cm. The apex, V, is over the centre of the rectangular base.

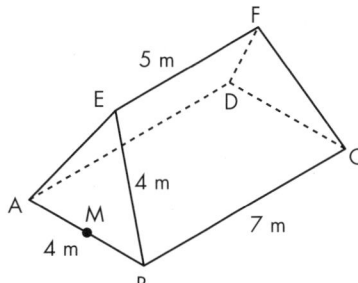

Calculate:

a) the length of the diagonal AC

b) the size of the angle VAC

c) the height of the pyramid.

3. The diagram shows the roof of a building. The base ABCD is a horizontal rectangle 7 m by 4 m. The triangular ends are equilateral triangles. Each side of the roof is an isosceles trapezium. The length of the top of the roof, EF, is 5 m.

Calculate:

a) the length EM, where M is the midpoint of AB

b) the size of angle EBC

c) the size of the angle between planes EAB and the base ABCD (the angle between EM and ABCD).

4. In the diagram, XABCD is a pyramid with a rectangular base.

a) Revina says that the angle between the edge XD and the base ABCD is 56.3°. Work out the correct answer to show that Revina is wrong.

b) Work out the angle between the planes ABCD and XDC.

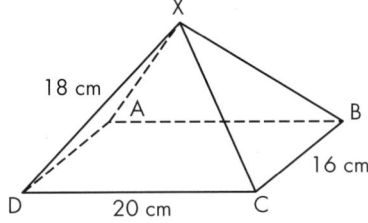

HINTS AND TIPS
Always write down intermediate working values to at least 4 significant figures, or use the entire answer on your calculator display to avoid inaccuracy in the final answer.

7.3 Sine rule, cosine rule and area of a triangle in 2D and 3D

THIS SECTION WILL SHOW YOU HOW TO ...
✓ understand and use the formulae for sine rule and cosine rule and area of a triangle in 2D and 3D
✓ prove the exact values of trigonometrical ratios for angles 30°, 45° and 60°

KEY WORDS
✓ sine rule   ✓ cosine rule   ✓ included angle   ✓ obtuse

The sine rule

Take a triangle ABC and draw the perpendicular from A to the opposite side BC.

From right-angled triangle ADB, $h = c \sin B$

From right-angled triangle ADC, $h = b \sin C$

Therefore,

$c \sin B = b \sin C$

which can be rearranged to give $\dfrac{c}{\sin C} = \dfrac{b}{\sin B}$

By drawing a perpendicular from each of the other two vertices to the opposite side (or by algebraic symmetry), you can show that

$\dfrac{a}{\sin A} = \dfrac{c}{\sin C}$ and $\dfrac{a}{\sin A} = \dfrac{b}{\sin B}$

These are usually combined to give the **sine rule**:

$\dfrac{a}{\sin A} = \dfrac{b}{\sin B} = \dfrac{c}{\sin C}$

The rule can be inverted to give

$\dfrac{\sin A}{a} = \dfrac{\sin B}{b} = \dfrac{\sin C}{c}$

Note:

- When you are calculating a *side*, use the rule with the *sides on top*.
- When you are calculating an *angle*, use the rule with the *sines on top*.

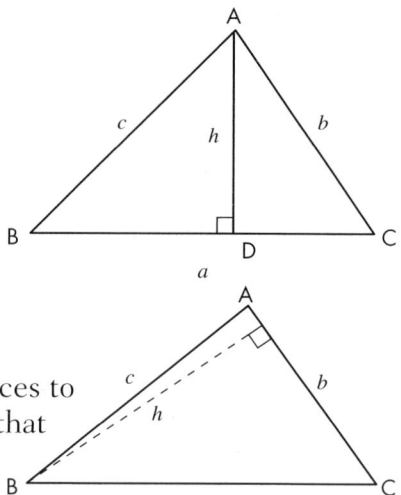

> **HINTS AND TIPS**
>
> When using the sine rule: take each side in turn, divide it by the sine of the angle opposite and then equate the results.

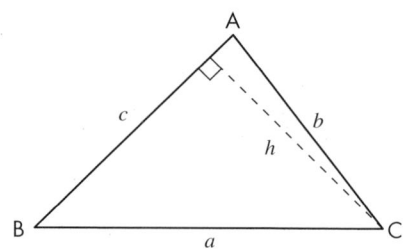

EXAMPLE 6

In triangle ABC, find the value of x. Give your answer to 3 significant figures.

Use the sine rule with sides on top. This gives

$\dfrac{x}{\sin 84°} = \dfrac{25}{\sin 47°}$

$\Rightarrow x = \dfrac{25 \sin 84°}{\sin 47°} = 34.0 \text{ cm (3 sf)}$

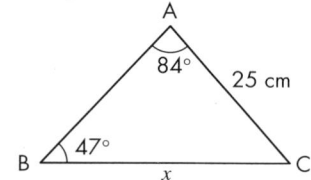

EXAMPLE 7

In the triangle ABC, find the value of the acute angle x. Give your answer to 1 decimal place.

Use the sine rule with sines on top, which gives:

$\dfrac{\sin x}{7} = \dfrac{\sin 40°}{6} \Rightarrow \sin x = \dfrac{7 \sin 40°}{6} = 0.7499$

$x = \sin^{-1} 0.7499$

$= 48.6° \text{ (1 dp)}$

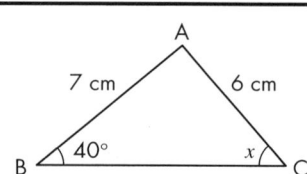

> **HINTS AND TIPS**
>
> The sine rule works even if the triangle has an obtuse angle, because you can find the sine of an obtuse angle.

EXERCISE 7E

TARGETING PASS

1. In each triangle, find the length of the side marked x. Give your answers to 3 significant figures.

a) 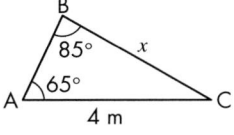   b) 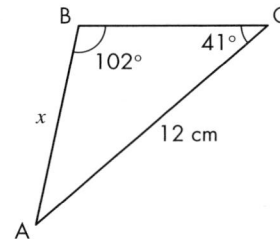   c)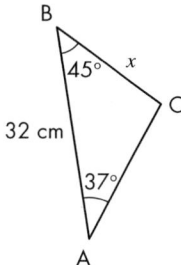

2. In each triangle, find the size of the angle marked x. Give your answers to 1 decimal place.

a) 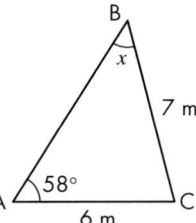   b) 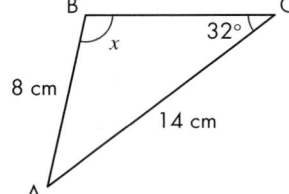   c)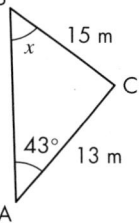

TARGETING MERIT

3. A mass is hung from a horizontal beam by two strings. The shorter string is 2.5 m long and makes an angle of 71° with the horizontal. The longer string makes an angle of 43° with the horizontal. Find the length of the longer string.

4. An old building is unsafe and is protected by a fence.

    A company is going to demolish the building and needs to work out its height BD, marked h on the diagram.

    Use the given information to calculate the value of h.

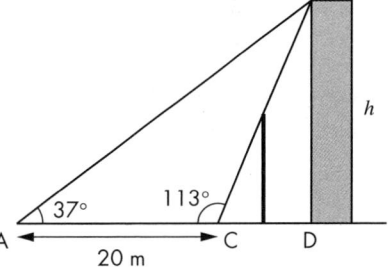

5. A rescue helicopter is based at an airfield at A. It is sent out to rescue a man from a mountain at M, due north of A.

    The helicopter then flies on a bearing of 145° to a hospital at H, as shown on the diagram.

    Calculate the direct distance from the mountain to the hospital.

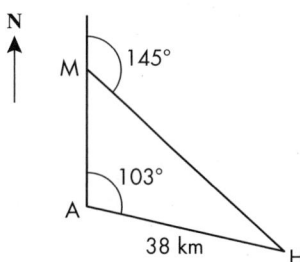

TARGETING DISTINCTION

6. A cube has edges of length 12 cm. A, B, H and J are vertices and P is a point on AB such that AP : PB = 2 : 1

a) Work out the distance PJ.

b) Work out the distance HJ.

c) Given that angle JHP = 50°, work out the size of angle JPH.

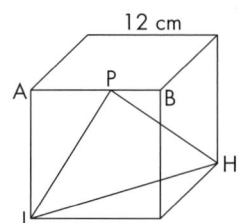

The cosine rule

In triangle ABC, D is the foot of the perpendicular to BC from A.
Using Pythagoras' theorem in triangle BDA, $h^2 = c^2 - x^2$

Using Pythagoras' theorem in triangle ADC, $h^2 = b^2 - (a - x)^2$
Therefore,
$c^2 - x^2 = b^2 - (a - x)^2 = b^2 - a^2 + 2ax - x^2 \Rightarrow c^2 = b^2 - a^2 + 2ax$

From triangle BDA, $x = c \cos B$.

$c^2 = b^2 - a^2 + 2ac \cos B \Rightarrow b^2 = a^2 + c^2 - 2ac \cos B$

By algebraic symmetry,

$a^2 = b^2 + c^2 - 2bc \cos A$ and $c^2 = a^2 + b^2 - 2ab \cos C$

This is the **cosine rule**. You can use the cosine rule when you are given two sides and the **included angle** (the angle between the given sides).

The formula can be rearranged to find any angle.

$$\cos A = \frac{b^2 + c^2 - a^2}{2bc} \qquad \cos B = \frac{a^2 + c^2 - b^2}{2ac} \qquad \cos C = \frac{a^2 + b^2 - c^2}{2ab}$$

EXAMPLE 8

Find the value of x in this triangle. Give your answer to 3 significant figures.
By the cosine rule:
$x^2 = 6^2 + 10^2 - 2 \times 6 \times 10 \times \cos 80° = 115.16$
so $x = \sqrt{115.16} = 10.7$
(3 significant figures)

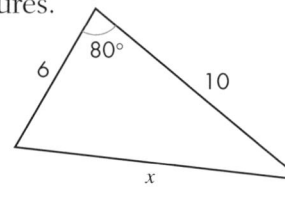

EXAMPLE 9

Find the value of x in this triangle. Give your answer correct to 1 decimal place.
By the cosine rule:
$\cos x = \frac{5^2 + 7^2 - 8^2}{2 \times 5 \times 7} = 0.1428 \Rightarrow x = 81.8°$ (1 dp)

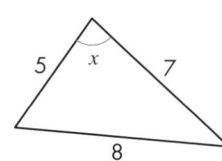

EXERCISE 7F

TARGETING PASS

1. In each triangle, find the length of the side marked x. Give your answers to 3 significant figures.

   a) 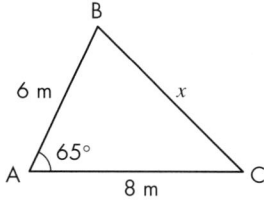 b) 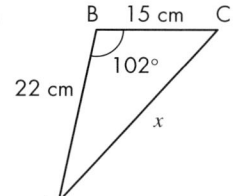 c)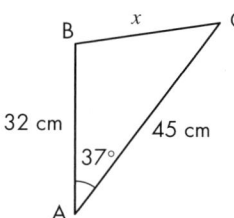

2. a) In each triangle, find the size of the angle marked x. Give your answers to 1 decimal place.

   i) 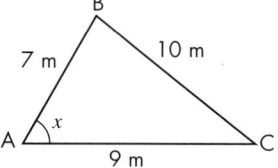 ii) 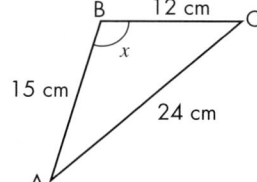 iii)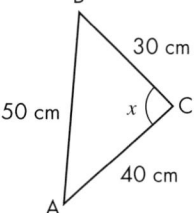

   b) Explain the significance of the answer to part **a iii**.

TARGETING MERIT

3. The diagram shows a trapezium ABCD. AB = 6.7 cm, AD = 7.2 cm, CB = 9.3 cm and angle DAB = 100°. Calculate:

   a) the length DB b) angle DBA
   c) angle DBC d) the length DC.

   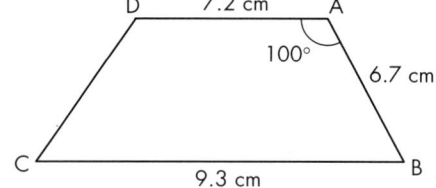

4. A ship sails from a port on a bearing of 050° for 50 km then turns onto a bearing of 150° for 40 km. A crewman is taken ill, so the ship drops anchor. What course and distance should a rescue helicopter from the port fly to reach the ship in the shortest possible time?

5. The three sides of a triangle are given as $3a$, $5a$ and $7a$. Calculate the smallest angle in the triangle.

6. Calculate the size of the largest angle in the triangle ABC.

   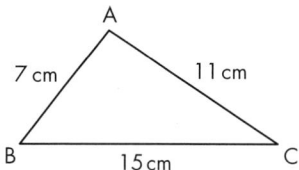

TARGETING DISTINCTION

7. In the rectangular based pyramid VABCD, AB = 8 cm, BC = 6 cm, VA = 12 cm and VC = 7 cm

   Work out the size of angle AVC.

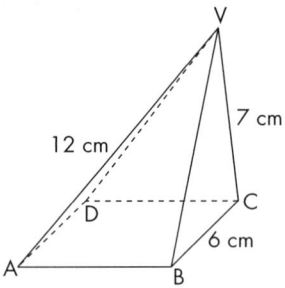

Choosing the correct rule

When you are solving a triangle, by calculating unknown measurements, there are three sets of information you may be given.

Two sides and the included angle

1. Use the cosine rule to find the third side.
2. Use the sine rule to find either of the other angles.
3. Use the sum of the angles in a triangle to find the third angle.

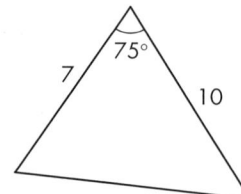

Two angles and a side

1. Use the sum of the angles in a triangle to find the third angle.

2, 3. Use the sine rule to find the other two sides.

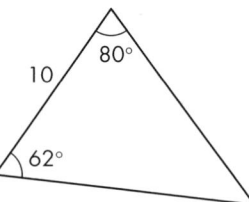

Three sides

1. Use the cosine rule to find one angle.
2. Use the sine rule to find another angle.
3. Use the sum of the angles in a triangle to find the third angle.

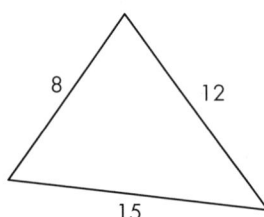

EXERCISE 7G

TARGETING PASS

1. In each triangle, find the value of the length or angle marked x. Give your answers to 3 significant figures.

a)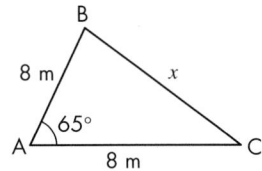

b)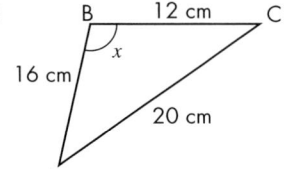

c)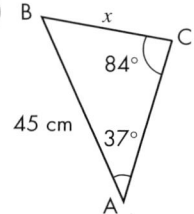

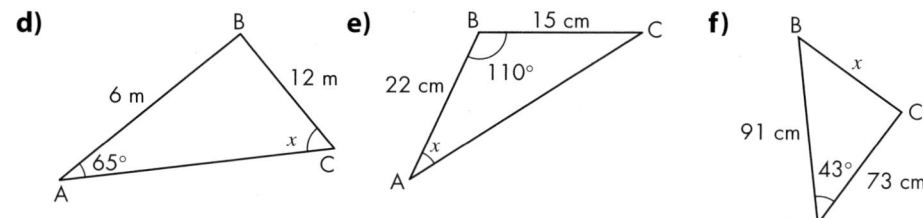

2. The hands of a clock have lengths 3 cm and 5 cm. Find the distance between the tips of the hands at 4 o'clock.

3. A spacecraft is seen hovering at a point which is in the same vertical plane as two towns, X and F, which are on the same level. Its distances from X and F are 8.5 km and 12 km, respectively. The angle of elevation of the spacecraft when observed from F is 43°.

   Calculate the distance between the two towns.

TARGETING MERIT

4. Triangle ABC has sides with lengths a, b and c, as shown in the diagram.
   a) What can you say about the angle BAC, if $b^2 + c^2 - a^2 = 0$?
   b) What can you say about the angle BAC, if $b^2 + c^2 - a^2 > 0$?
   c) What can you say about the angle BAC, if $b^2 + c^2 - a^2 < 0$?

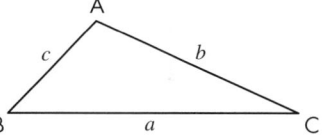

5. The diagram shows a sketch of a field ABCD.
   A farmer wants to put a new fence round the perimeter of the field.
   Calculate the perimeter of the field.

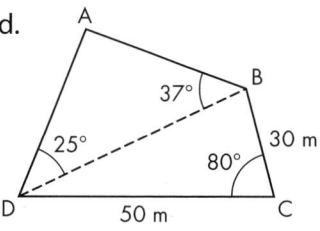

TARGETING DISTINCTION

6. A ship sails due north from port P for 119 km to reach port Q.

   It then sails 78 km on a bearing of 125° to another port R.

   Work out the bearing of P from R, giving your answer to the nearest degree.

TARGETING DISTINCTION*

7. A triangle has two sides of lengths 2 and $\sqrt{2}$. The area of the triangle is 1 cm².

   Find the exact length of the third side of the triangle and the interior angles of the triangle.

8. Three points $X(6, -1)$, $Y(1, -4)$ and $Z(4, 1)$ are joined to form a triangle.

   Calculate the area of the triangle.

Proof of the exact values of trigonometrical ratios for angles 30°, 45° and 60°

An equilateral triangle can be divided into two right-angled triangles.
Let the sides of the equilateral triangle have length 2 units, as shown.
You can use Pythagoras' theorem to work out the height of the triangle.
$h^2 = 2^2 - 1^2 = 3 \Rightarrow h = \sqrt{3}$ units

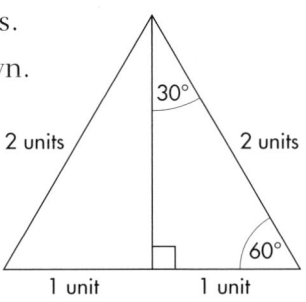

Then one of the right-angled triangles looks like this.
Using trigonometrical ratios:
$\sin x = \dfrac{\text{opposite}}{\text{hypotenuse}}$, $\cos x = \dfrac{\text{adjacent}}{\text{hypotenuse}}$ and $\tan x = \dfrac{\text{opposite}}{\text{adjacent}}$

This gives
$\sin 30° = \dfrac{1}{2}$, $\cos 30° = \dfrac{\sqrt{3}}{2}$, $\tan 30° = \dfrac{1}{\sqrt{3}}$
$\sin 60° = \dfrac{\sqrt{3}}{2}$, $\cos 60° = \dfrac{1}{2}$, $\tan 60° = \sqrt{3}$

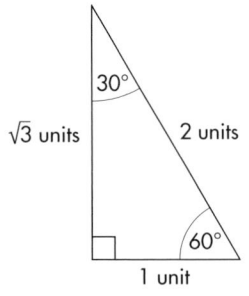

An isosceles right-angled triangle has angles of 45°, 45° and 90°.
Let the sides forming the right angle have lengths 1 unit, as shown.
You can use Pythagoras' theorem to work out the length of the hypotenuse.
$x^2 = 1^2 + 1^2 = 2$
$x = \sqrt{2}$ units

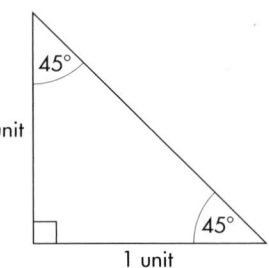

Using trigonometrical ratios:
$\sin x = \dfrac{\text{opposite}}{\text{hypotenuse}}$, $\cos x = \dfrac{\text{adjacent}}{\text{hypotenuse}}$ and $\tan x = \dfrac{\text{opposite}}{\text{adjacent}}$

This gives
$\sin 45° = \dfrac{1}{\sqrt{2}}$, $\cos 45° = \dfrac{1}{\sqrt{2}}$, $\tan 45° = 1$

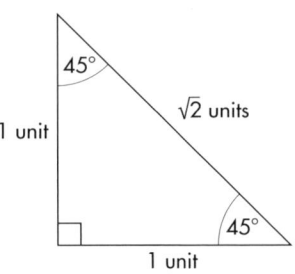

You will need to learn these ratios. Use this table to help you.

	sin	cos	tan
30°	$\dfrac{1}{2}$	$\dfrac{\sqrt{3}}{2}$	$\dfrac{1}{\sqrt{3}}$
45°	$\dfrac{1}{\sqrt{2}}$	$\dfrac{1}{\sqrt{2}}$	1
60°	$\dfrac{\sqrt{3}}{2}$	$\dfrac{1}{2}$	$\sqrt{3}$

EXAMPLE 10

Work out the exact value of x in this triangle.
Using $\sin 30° = \dfrac{\sqrt{3}}{x}$ gives $\dfrac{1}{2} = \dfrac{\sqrt{3}}{x}$.
Rearranging gives $x = 2\sqrt{3}$ cm.

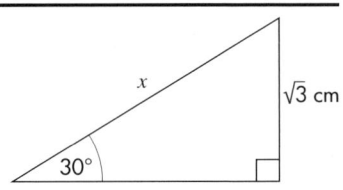

EXERCISE 7H

TARGETING PASS

In this exercise, leave your answers in surd form.

1. Work out the value of x in each triangle.

a)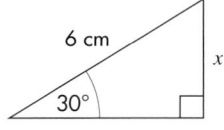

b)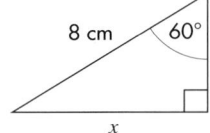

c)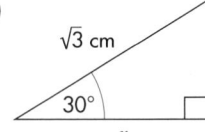

d)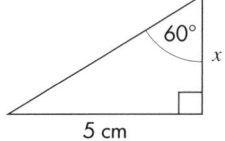

e)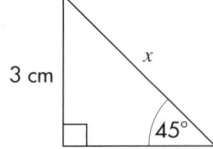

f)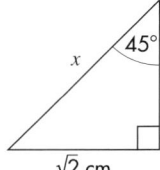

TARGETING MERIT

2. Work out the value of x in each diagram.

a)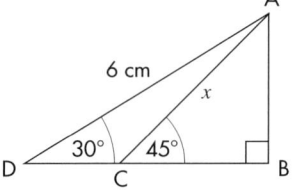

b)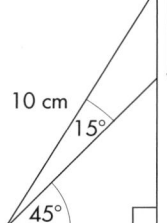

c)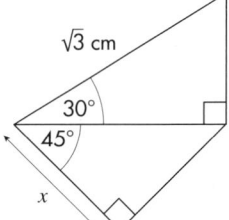

How to solve problems involving area of a triangle

The area of a triangle $= \frac{1}{2}ab\sin C$

You can use this formula when you do not know the height of the triangle, but you are given the lengths of two sides and the size of the angle between them.

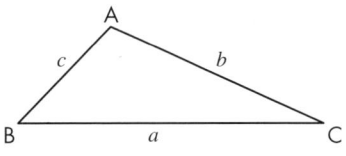

EXAMPLE 11

The area of this triangle is 56 cm².
Work out the size of angle X.
Give your answer to one decimal place.

$\frac{1}{2}ab\sin C = \text{area}$

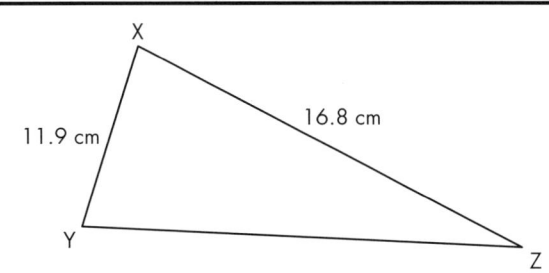

Substitute the lengths from the diagram into the formula and make it equal to the area, which is given as 56 cm²

$$\tfrac{1}{2} \times 11.9 \times 16.8 \times \sin X = 56$$

Make sin X the subject by rearranging:

$$\sin X = \frac{56 \times 2}{11.9 \times 16.8}$$

Solve by taking the inverse sine of both sides:

$$X = \sin^{-1} \frac{112}{199.92}$$
$$= 34.1° \text{ (to 1 dp)}$$

EXAMPLE 12

A square-based pyramid has a base length of 7 cm.

Triangle ACD is isosceles.

Angle ACD is 63°.

Work out the surface area of the pyramid. Give your answer to two decimal places.

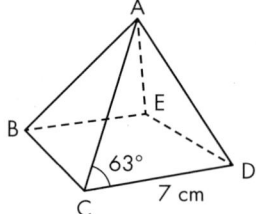

Triangle ACD is isosceles and so angle ADC equals 63° and angle CAD = 180 − 63 − 63 = 54°

Using the sine rule $\dfrac{7}{\sin 54°} = \dfrac{AD}{\sin 63°}$

$$AD = \frac{7 \times \sin 63°}{\sin 54°} = 7.70941......$$

Area of triangle ACD = $\tfrac{1}{2} \times 7 \times 7.70941.... \times \sin 63° = 24.0419....$

Surface area = $(7 \times 7) + (4 \times 24.0419...) = 145.17$ cm² (to 2 dp)

Exam-style questions

1. In triangle ABC, the angle ADC is a right angle.

   AB = 13 cm, AD = 12 cm and AC = 15 cm.

   Work out the area of triangle ABC.

   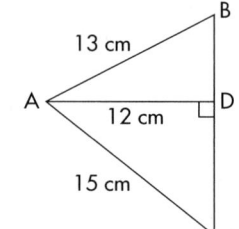

   [4 marks]

2. ABD and DBC are right-angled triangles.

   Angle ABD = angle DBC = 30°.

   The length of BC is 6 cm.

   Work out the length of AB.

   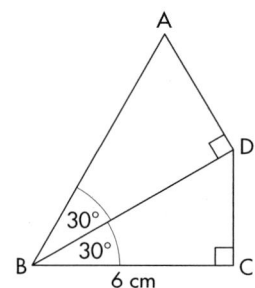

   [4 marks]

3. In triangle ABC, AC = AB = 4 cm and CB = 5 cm.

   Work out the exact value of cos A. [2 marks]

   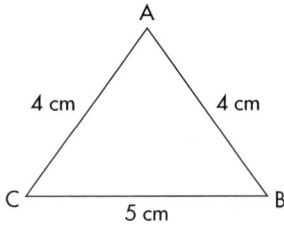

4. In triangle ABC, AB = 5 cm, BC = 6 cm and angle ABC = 55°. Find the length of AC. [3 marks]

5. The diagram shows an equilateral triangle of side 15 cm with a circle passing through the three vertices.

   Calculate the diameter of the circle. [4 marks]

   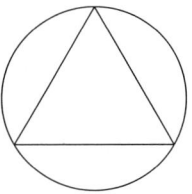

6. ABCD is the top face of a cuboid and EFGH is the base.

   The lengths of the sides of the cuboid are 6 cm, 8 cm and 10 cm.

   M is the midpoint of AB.

   a) Calculate the length of MH. [4 marks]
   b) Calculate the angle between MH and the base EFGH. [3 marks]
   c) Calculate the angle between triangle MGH and face CDHG. [3 marks]

   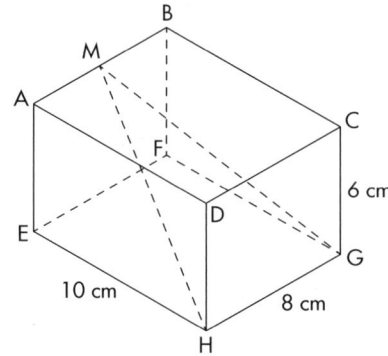

7. In pyramid VABCD, VA = VB = VC = VD = 9 cm M is the midpoint of DC.

   a) Show that VM = $\sqrt{65}$ cm. [2 marks]
   b) Work out the size of angle VAM. [4 marks]

   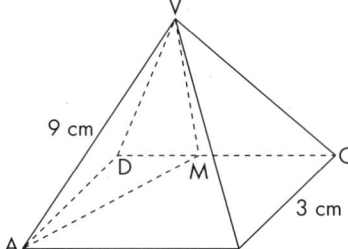

8. The diagram shows a regular pentagon inscribed in a circle with radius 9 cm.

   Work out the area of the pentagon, giving your answer to three significant figures. [4 marks]

   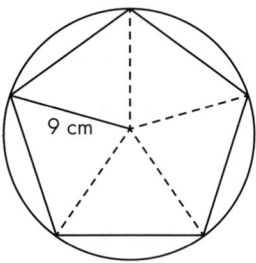

9. A wooden doorstop is a triangular based prism and is shown in the diagram below.

   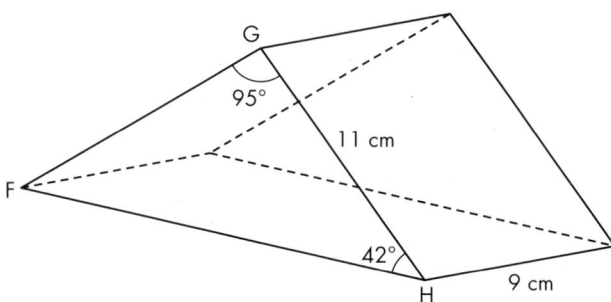

   Work out the volume of the doorstop. Give your answer to three significant figures. [4 marks]

10. Find the value of x^2, giving your answer in the form $a + b\sqrt{7}$. [4 marks]

    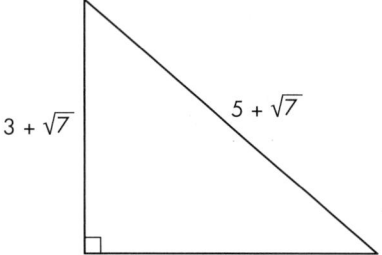

8 Probability

8.1 The language of probability

THIS SECTION WILL SHOW YOU HOW TO ...
✓ understand and use the language and notation of probability
✓ understand independent and dependent events
✓ understand and solve problems with enumerate sets of probabilities
✓ understand and solve problems with combinations of sets of probabilities

KEY WORDS
✓ independent and dependent events
✓ complement
✓ subset
✓ empty set
✓ sample space
✓ mutually exclusive
✓ universal set

Independent and dependent events

Two events are called **independent** if the probability of one event occurring is unaffected by the occurrence of the other event. Therefore, if A and B are independent, the probability of A happening is the same, whether or not B happens.

The shaded region in the diagram shows the 'intersection' of A and B, which is denoted by $A \cap B$. This represents A and B both happening.

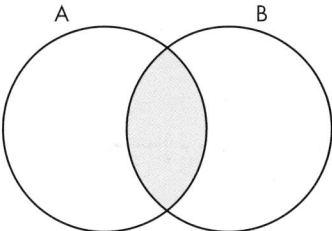

A and B are independent if:

$P(A \cap B) = P(A) \times P(B)$

where $P(A \cap B)$ is the probability of A *and* B happening, P(A) is the probability of A happening and P(B) is the probability of B happening.

This implies that:

$P(A \mid B)$ (the probability of A given B has happened) $= P(A)$

and

$P(B \mid A)$ (the probability of B given A has happened) $= P(B)$

Rearranging the equation gives

$P(A \cap B) = P(A \mid B) \times P(B)$ and $P(A \cap B) = P(A) \times P(B \mid A)$

Also, $P(A \cap B) = P(B \cap A)$, which means that the probability represented by the intersection of A and B is the same as the probability represented by the intersection of B and A

Two events are **dependent** if the probability of one event happening is affected by the other event occurring (or not occurring).

If probabilities are affected by previous events, then they *cannot* be independent.

Taking a counter from a bag and not replacing it and then taking another counter would *not* be an example of two independent events.

EXAMPLE 1

In a game, the player first has to flip a coin and then spins a 6-sided spinner. The player wins a prize if the coin shows tails and the score on the spinner is 5 or 6. Find the probability that the player wins the prize.

The question can be answered by listing out all the outcomes in the **sample space**, that is, {(H,1), (H,2), (H,3), (H,4), (H,5), (H,6), (T,1), (T,2), (T,3), (T,4), (T,5), (T,6)}, but it is more efficient to use the fact that the events are independent.

Therefore, $P(T \cap 5 \text{ or } 6) = P(T) \times P(5 \text{ or } 6)$
$$= \tfrac{1}{2} \times \tfrac{2}{6} = \tfrac{2}{12} = \tfrac{1}{6}$$

> **HINTS AND TIPS**
>
> Many questions can be solved much more efficiently using properties of independent events than by considering the sample space.

EXAMPLE 2

In an experiment, a bag contains 5 green counters and 6 yellow counters. One counter is selected at random from the bag and then replaced before a second counter is randomly selected. Find the probability that:

a) both counters are green **b)** both counters are yellow

c) only one counter is green **d)** at least one counter is yellow

e) the second counter is yellow.

Remember that the probability of an event happening is

$$\frac{\text{number of outcomes of the event}}{\text{total number of outcomes}}$$

So, $P(\text{Green}) = \tfrac{5}{11}$ and $P(\text{Yellow}) = \tfrac{6}{11}$

Draw a tree diagram.

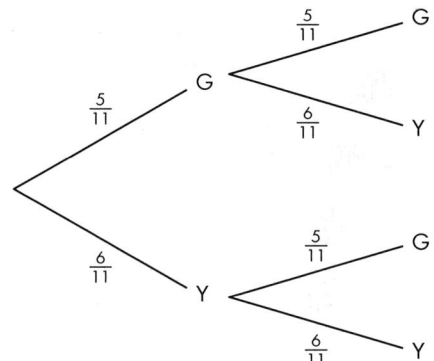

a) Multiply along the branches to find the probability of selecting G and G.

The probability that both counters are green

$P(G,G) = \tfrac{5}{11} \times \tfrac{5}{11} = \tfrac{25}{121}$

b) Similarly, the probability that both are yellow

$P(Y,Y) = \tfrac{6}{11} \times \tfrac{6}{11} = \tfrac{36}{121}$

c) The possible combinations are G and Y or Y and G. Add the probabilities of these combinations.

The probability that only one counter is green is given by

$P(G,Y) + P(Y,G) = \left(\tfrac{5}{11} \times \tfrac{6}{11}\right) + \left(\tfrac{6}{11} \times \tfrac{5}{11}\right) = \tfrac{60}{121}$

d) The probability that at least one counter is yellow (which is the same as every option apart from Green, Green) can be calculated using

$1 - P(G,G) = 1 - \left(\tfrac{5}{11} \times \tfrac{5}{11}\right) = 1 - \tfrac{25}{121} = \tfrac{96}{121}$

e) The probability that the second counter is yellow is given by

$P(G,Y) + P(Y,Y) = \left(\tfrac{5}{11} \times \tfrac{6}{11}\right) + \left(\tfrac{6}{11} \times \tfrac{6}{11}\right) = \tfrac{66}{121}$

More language and notation of probability

P(A′) means the probability of 'not' A. This is also called the **complement** of A.
This can be calculated using P(A′) = 1 − P(A).

The shading in the following diagram represents A ∪ B, which denotes the set that consists of all elements belonging to either set A or set B (or both).

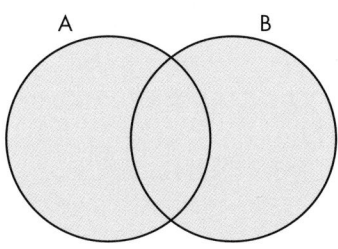

P(A ∪ B) = P(A) + P(B) − P(A ∩ B)

Two events are **mutually exclusive** if the occurrence of one of them prevents the occurrence of the other. In other words, they cannot both happen at the same time.

For example, if event A is 'passing the exam' and event B is 'not passing the same exam', then they cannot occur at the same time and are therefore mutually exclusive events

Using set notation, you can say that A intersection B is an **empty set**
A ∩ B = ∅ and also that P(A ∩ B) = 0

It is also true that if A and B are mutually exclusive events, then
P(A ∪ B) = P(A) + P(B).

In the diagram, S represents the sample space for all the possible outcomes of an experiment. It shows that A and B are **subsets** of S, which can be denoted by A ⊂ S and B ⊂ S, respectively.

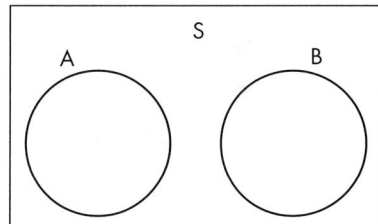

The sets A and B do not intersect, which shows that events A and B are mutually exclusive.

EXAMPLE 3

The Venn diagram shows three events A, B and C.

ξ is the **universal set** (the set containing all objects or elements and of which all other sets are subsets).

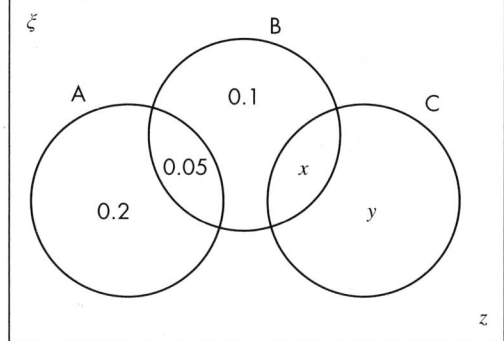

a) Which two events are mutually exclusive?

Events A and B are independent. P(C) = 0.4

b) Find the values of x, y and z.

c) Find P(A′)

a) A and C are mutually exclusive as they do not intersect.

b) If A and B are independent, then P(A ∩ B) = P(A) × P(B)

$\quad$ P(A ∩ B) = 0.05 (from the diagram) $\Rightarrow$ 0.05 = (0.2 + 0.05) × (0.1 + 0.05 + x)

$$0.05 = 0.25 \times (0.15 + x)$$
$$0.2 = 0.15 + x$$
$$0.05 = x \Rightarrow x = 0.05$$
$$P(C) = 0.4 = x + y \Rightarrow 0.4 = 0.05 + y$$
$$y = 0.35$$

$$0.2 + 0.05 + 0.1 + x + y + z = 1 \Rightarrow z = 1 - 0.35 - 0.05 - 0.35$$
$$z = 0.25$$

c) $P(A) = 0.25 \Rightarrow P(A') = 1 - 0.25 = 0.75$

EXERCISE 8A

TARGETING PASS

1. In an experiment a bag contains 4 blue counters and 6 red counters. One counter is selected at random from the bag and then replaced before a second counter is randomly selected. Find the probability that:

   a) both counters are blue
   b) both are red
   c) only one counter is blue
   d) at least one counter is red
   e) the second counter is green.

   In a second experiment using the same bag, a counter is selected at random and not replaced.

   f) Does the fact that the counter is not replaced make a difference to your answers? If so, explain why.

   HINTS AND TIPS

   In probability questions you are not expected to cancel or simplify the fractions. It is easier to leave the denominators the same in case you need to do further calculations.

EXAMPLE 4

A tin of biscuits contains only digestives and shortbread.

There are 5 digestive biscuits.

Mina takes out one biscuit and replaces it before taking another.

The probability that Mina takes out the same type of biscuit is $\frac{17}{32}$.

Using algebra, work out the number of shortbread biscuits in the tin.

First, draw a tree diagram to represent the outcome algebraically. Let the total number of biscuits be x.

Then form an equation: $\quad P(D, D) + P(S, S) = \frac{17}{32}$

$$\left(\frac{5}{x} \times \frac{5}{x}\right) + \left(\frac{x-5}{x} \times \frac{x-5}{x}\right) = \frac{17}{32}$$

Expand the brackets: $\quad \dfrac{25}{x^2} + \dfrac{x^2 - 10x + 25}{x^2} = \dfrac{17}{32}$

Simplify: $\quad \dfrac{x^2 - 10x + 50}{x^2} = \dfrac{17}{32}$

Cross multiply: $\quad 32(x^2 - 10x + 50) = 17x^2$

Form a quadratic equation: $\quad 32x^2 - 320x + 1600 = 17x^2$

$$15x^2 - 320x + 1600 = 0$$

Divide every term by 5: $\qquad 3x^2 - 64x + 320 = 0$
Solve by either factorising or
by using the quadratic formula: $\quad (3x - 40)(x - 8) = 0$
$$x = \tfrac{40}{3} \text{ or } x = 8$$

x must be an integer and so there $8 - 5 = 3$ shortbread biscuits in the tin.

EXERCISE 8B

TARGETING MERIT

1. A box of fruit contains only apples and bananas. There are 6 apples in the box.
   Sam takes out a piece of fruit, replaces it, and then takes out another piece of fruit.
   The probability that Sam takes out one of each fruit is $\frac{12}{25}$.
   Using algebra, work out the two possible numbers of bananas in the box.

TARGETING DISTINCTION

2. The events A and B are independent.
   For $P(A) = 0.6 \quad P(B) = 0.8 \quad$ calculate: **a)** $P(A \cap B)$ **b)** $P(A \mid B)$

3. C and D are two events
   $P(C) = 0.05 \quad P(D) = 0.15 \quad P(C \cap D) = 0.05$
   Determine whether or not C and D are independent.

4. E and F are two events
   $P(E) = 0.1 \quad P(F) = 0.3 \quad P(E \cap F) = 0.03$
   Determine whether or not E and F are independent.

5. $P(G) = \frac{2}{5} \quad P(H) = \frac{1}{5} \quad P(G \cap H) = \frac{2}{25}$
   Determine whether or not events G and H are independent.

6. $P(A') = \frac{3}{4} \quad P(B) = \frac{3}{10} \quad P(A \cap B) = \frac{3}{40}$
   Determine whether or not events A and B are independent.

For questions **7** to **9**, A and B are independent.

7. If $P(A) = 0.25$ and $P(B) = 0.6$, work out $P(B \mid A)$.
8. If $P(B) = \frac{9}{20}$ and $P(A \cap B) = \frac{9}{100}$, work out $P(A)$.
9. If $P(A) = 0.4$ and $P(A \cap B) = 0.3$, work out $P(B')$.

HINTS AND TIPS

In question 5, $P(A')$ means the probability of *not* A (or the complement of A).
A and B are independent if $P(A \cap B)$ (the probability of A *and* B happening) = $P(A) \times P(B)$
Remember these formulae:
$P(A \cap B) = P(A \mid B) \times P(B)$ and
$P(A \cap B) = P(A) \times P(B \mid A)$

TARGETING DISTINCTION*

10. A, B and C are three events: P(A) = 0.25, P(B) = 0.3, P(C) = 0.45 and P(B ∩ C) = 0.1

   a) Given that the events A and C are mutually exclusive, and that the events A and B are independent, draw a Venn diagram to show all the probabilities.

   b) Find:

      i) P(A ∩ B)

      ii) P(A ∩ C)

      iii) P(A ∪ B ∪ C)′

      iv) P(A′ ∩ C′)

      v) P((A ∩ B′) ∪ C)

11. The independent events, A and B, are such that $P(A) = \frac{1}{3}$, $P(B) = \frac{1}{4}$ and $P(A \cap B) = x$.

   a) Work out the value of x.

   b) Represent these probabilities on a Venn diagram.

   c) Find P(A′ ∩ B′).

8.2 Conditional probability

THIS SECTION WILL SHOW YOU HOW TO ...
✓ use two-way tables, Venn diagrams and tree diagrams to solve conditional probability problems

KEY WORDS
✓ conditional probability
✓ two-way table

When an event is dependent on a previous event, then the probability is **conditional**.

This means that the probability of an event happening will depend on what has occurred before.

EXAMPLE 5

Two beads are selected at random from a bag that contains 3 blue, 2 red and 6 green beads.
Find the probability that both beads are different colours.

You can draw a tree diagram to help with this question (although you could calculate the answer without the diagram).

You could answer the question by finding the probabilities of all the combinations of coloured beads for which the colours are different (there are 6 possible combinations) and then adding these together. However, it is more efficient to find the sum of the probabilities of B,B, R,R and G,G (that is, the probabilities that the colours are the same) and subtract the sum from 1 (this leaves all the combinations where the colours are different).

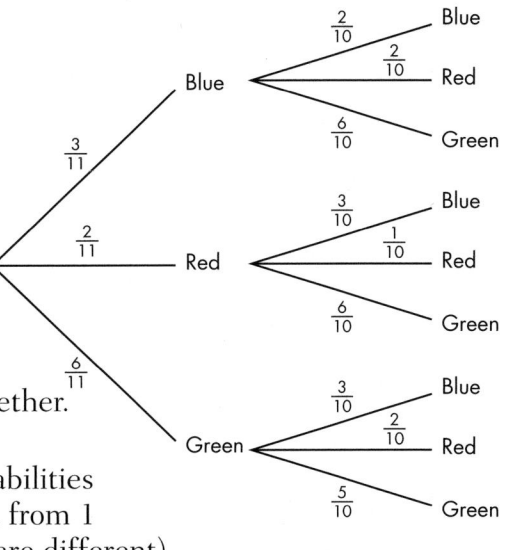

P(two different colours) = 1 − P(same colour) = 1 − (P(B,B) + P(R,R) + P(G,G))

$$= 1 - \left(\left(\frac{3}{11} \times \frac{2}{10}\right) + \left(\frac{2}{11} \times \frac{1}{10}\right) + \left(\frac{6}{11} \times \frac{5}{10}\right)\right)$$

$$= 1 - \left(\frac{6}{110} + \frac{2}{110} + \frac{30}{110}\right)$$

$$= 1 - \left(\frac{38}{110}\right)$$

$$= \frac{72}{110} \text{ (You do not need to cancel the fraction.)}$$

EXAMPLE 6

One week, 200 children tried a new sports activity.
Each child chose football, tennis or swimming.

15% of the children chose swimming
The ratio of children who tried weekend football to weekend tennis was 1 : 2.
81 children tried tennis.
21 children tried weekend swimming.
83 children tried a sport on a weekday.

a) Work out the probability that a child, chosen at random, tried weekday football.
b) Given that a child tried a sport at the weekend, what is the proability that they chose tennis?

Draw a **two-way table** to answer this question.

	Football	Tennis	Swimming	Total
Weekday	83 − 17 − 9 = **57**	81 − 64 = **17**	30 − 21 = **9**	83
Weekend	(117 − 21) ÷ 3 = **32**	(117 − 21) ÷ 3 × 2 = **64**	21	200 − 83 = **117**
Total	200 − 30 − 81 = **89**	81	15% of 200 = **30**	200

a) The probability that a child, chosen at random, tried weekday football = $\frac{57}{200}$.

b) The probability that a child, given they tried a sport at the weekend, chose tennis = $\frac{64}{117}$.

HINTS AND TIPS
Notice that the denominator in **b** is 117, not 200 as in **a**, because it is given that the child tried a sport at the weekend, so this is the total number of children to use when writing the probability.

EXERCISE 8C

TARGETING PASS

1. There are 250 shapes in a bag.

    18% of the shapes are red triangles

    The ratio of blue circles to blue squares to blue triangles is 2 : 4 : 1

    There are 131 red shapes

    There are 8 more red triangles than red circles.

    a) Copy and complete the two-way table.

	Circle	Square	Triangle	Total
Red				
Blue				
Total				

    b) Work out the probability that a shape, chosen at random, is a blue circle.

    c) Given that a shape is a square, what is the proability that it is red?

TARGETING MERIT

2. A bag contains 9 purple counters and 6 white counters.

    A counter is taken at random from the bag and **not replaced**.

    A second counter is then removed from the bag.

a) Draw a tree diagram to represent this information.

b) Find the probability that:
   i) both counters are white
   ii) both counters are purple
   iii) there is one counter of each colour
   iv) there is at least one white counter.

3. The probability that it will rain on a day in December is 0.17.
   When it rains, the probability that my bus is late is 0.63.
   When it does not rain, the probability that my bus is **not** late is 0.94.
   a) Draw a tree diagram to show this information.
   b) Calculate the probability that, on a day in December, it does not rain and my bus is late.
   c) Calculate the probability that the bus is not late.

4. Fay 30 has biscuits in a tin.
   There are: 15 plain biscuits
              8 chocolate biscuits
              7 coconut biscuits.
   Fay takes at random two biscuits from the tin and eats them.
   Calculate the probability that the biscuits are of the same type.

EXAMPLE 7

A and B are two events of the sample space where
$P(A) = 0.4$, $P(B) = 0.75$ and $P(B \mid A) = 0.675$

Find:
a) $P(A \cap B)$
b) $P(A \cup B)$
c) $P(A \mid B)$
d) $P(B' \mid A)$
e) $P(A' \cap B')$

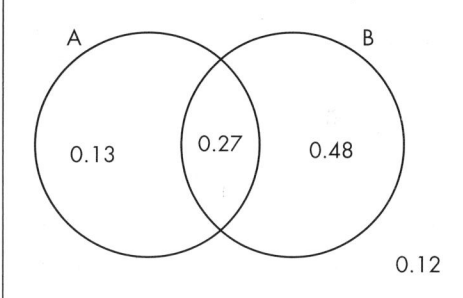

a) $P(B \mid A) = \dfrac{P(A \cap B)}{P(A)}$ and so $P(A \cap B) = P(B \mid A) \times P(A) = 0.675 \times 0.4 = 0.27$

b) $P(A \cup B) = P(A) + P(B) - P(A \cap B) = 0.4 + 0.75 - 0.27 = 0.88$

c) $P(A \mid B) = \dfrac{P(A \cap B)}{P(B)} = \dfrac{0.27}{0.75} = 0.36$

d) $P(B' \mid A) = \dfrac{P(B' \cap A)}{P(A)} = \dfrac{(P(A) - P(A \cap B))}{P(A)} = \dfrac{0.13}{0.4} = 0.325$

e) $P(A' \cap B') = 1 - P(A \cup B) = 1 - 0.88 = 0.12$

EXAMPLE 8

There are x chocolates in a box.
5 of the chocolates and dark and the rest are milk chocolate.
Mo takes two chocolates at random out of the box and eats them.
The probability that Mo eats two dark chocolates is $\frac{2}{11}$.
Work out the value of x.

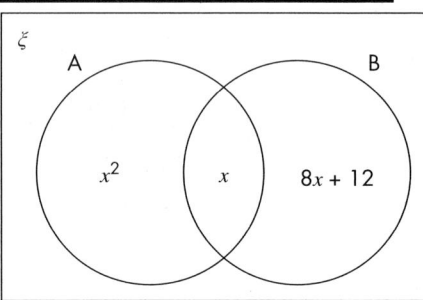

First draw a tree diagram to show all the outcomes:

Form an equation using the information given in the problem:

P(D, D) $\quad \frac{5}{x} \times \frac{4}{x-1} = \frac{2}{11}$

Simplify: $\quad \frac{20}{x(x-1)} = \frac{2}{11}$

Cross multiply: $\quad 20 \times 11 = 2x(x-1)$
$$220 = 2x^2 - 2x$$

Rearrange: $\quad 2x^2 - 2x - 220 = 0$

Divide by two: $\quad x^2 - x - 110 = 0$

Factorise: $\quad (x-11)(x+10) = 0$
$$x = 11 \text{ (reject } x = -10\text{)}$$

As x cannot be negative, you do not need to consider the other solution.

EXAMPLE 9

In the Venn diagram, $A \cup B = 64$, $x > 0$
Work out the value of x.

If $A \cup B = 64$, then you can form an equation using the expressions in the union of A and B.
$$x^2 + x + 8x + 12 = 64$$
$$x^2 + 9x - 52 = 0$$

Now solve this equation to find x: $\quad (x-4)(x+13) = 0$

As $x > 0$, $x = 4$

EXERCISE 8D

TARGETING DISTINCTION

1. F and G are two events of the sample space where $P(F) = \frac{1}{4}$, $P(G) = \frac{39}{100}$ and $P(G \mid F) = \frac{9}{25}$.
   Find:
   a) $P(F \cap G)$ b) $P(F \cup G)$ c) $P(F \mid G)$ d) $P(F' \mid G)$ e) $P(F' \cap G')$

2. Jay makes x lemon cakes, and 5 chocolate cakes and puts them on a tray
   He takes two cakes from the tray at random.

The probability that Jay takes one of each cake is $\frac{6}{11}$

a) Form a quadratic equation that models this problem.

b) Find the value of x.

3. Sasha has x red t-shirts, 4 blue t-shirts and 3 green t-shirts.

   Sasha takes two t-shirts at random from her drawer.

   The probability that Sasha takes two t-shirts that are the same colour is $\frac{3}{8}$.

   Find the total number of t-shirts that Sasha has.

TARGETING DISTINCTION*

4. In the Venn diagram, A = $\frac{1}{2}$B and (A ∪ B) = 60, $x > 0$

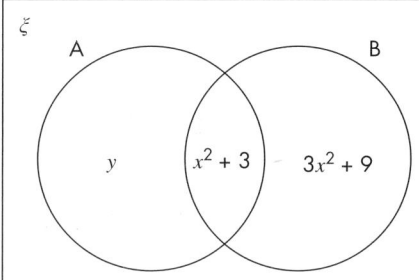

   Work out the value of x and the value of y.

5. There are 10 pencils in a case.

   There are x green pencils, and the rest of the pencils are black.

   Molly takes, at random, two pencils from the case.

   a) Draw a tree diagram to represent this information.

   b) Find an expression, in terms of x, for the probability that Molly takes one pencil of each colour. Simplify your answer if possible.

   c) Molly replaces the two pencils and then takes three pencils out of the case.

   The probability that Molly takes three black pencils is $\frac{1}{120}$.

   Find the value of x.

Exam-style questions

1. Gaby takes two tests, which she can either pass or fail.

   The probability of her passing the first test is 0.7 and the probability of her passing the second test is x.

   a) Draw a tree diagram to represent this. [2 marks]

   b) Write down **one** mathematical assumption that has been made about the two tests. [1 mark]

   c) The probability of Gaby passing only one of the two tests is 0.52.

   Work out the value of x. [4 marks]

2. If Rory goes to the gym on any given day, the probability that he will go on the next day is 0.3.

   If he does not go to the gym on a given day, the probability that he will go on the next day is 0.75.

   Rory went to the gym on Sunday.

   Find the probability that Rory went to the gym on the following Tuesday. [3 marks]

3. Abdi flips a biased coin.

   The probability that the coin will land on tails is 0.67.

   Abdi throws the coin three times.

   He says, 'The probability that the coin will land on tails all three times is less than 0.3.'

   Is Abdi correct? Explain your answer. [3 marks]

4. Dwayne plays three games, A, B and C.

   The Venn diagram below shows the probabilities of him winning each of the games.

   The probability that he wins at least one of the games is $\frac{47}{50}$ and the probability that he wins only one game is $\frac{31}{60}$.

   a) Work out the values of x, y and z. [4 marks]
   b) Show that A and B are independent events. [1 mark]

5. Tom plays a game that he can either win or lose. He wins the game more than he loses.

   Tom plays the game twice.

   The probability that he wins exactly one of the two games is 0.455.

   Work out the probability that Tom loses the next game he plays. [5 marks]

9 Proof

9.1 Proof by deduction

THIS SECTION WILL SHOW YOU HOW TO ...
✓ use algebra to support and construct proof by deduction

KEY WORDS
✓ proof   ✓ deduction   ✓ show   ✓ verify   ✓ identity

Proof by **deduction** is a method that starts with a theorem or a known fact and then uses logical steps to reach a final conclusion or statement showing that the theorem is true.

Here is a fact that you will already know.

The sum of any two odd numbers is always even

If you take any two odd numbers, add them together and get a number that can be divided exactly by 2. However, this does not prove the result, even if everyone in your class, or your school, or the whole country, did this for a different pair of starting odd numbers. Unless you tried every pair of odd numbers (and there is an infinite number of them) you cannot be 100% certain that this result is always true.

This is how to prove the result by deduction.

Let n be any whole number.

Whatever whole number is represented by n, $2n$ will always be even. So, $2n + 1$ represents any odd number.

Let one odd number be $2n + 1$, and let the other odd number be $2m + 1$.

The sum of these is

$(2n + 1) + (2m + 1) = 2n + 2m + 1 + 1 = 2n + 2m + 2 = 2(n + m + 1)$ which must be even.

This proves the result, as n and m can be any numbers.

In algebraic proof by deduction, you must show every step clearly and the algebra must be clear.

There are three levels of 'proof': **Verify** that ..., **Show** that ... and **Prove** that ...

- At the lowest level (verification), you just need to substitute numbers into the result to show that it works.
- At the middle level (show that), you have to show that both sides of the result are the same algebraically.
- At the highest level (proof), you have to manipulate the left-hand side (LHS) of the result to become its right-hand side (RHS).

The following example demonstrates these three different procedures.

EXAMPLE 1

You are given that $n^2 + (n + 1)^2 - (n + 2)^2 = (n - 3)(n + 1)$.

a) Verify that this result is true.
b) Show that this result is true.
c) Prove that this result is true.

a) Choose a number for n, say $n = 5$. Substitute this value into both sides of the expression:
$5^2 + (5 + 1)^2 - (5 + 2)^2 = (5 - 3)(5 + 1)$
$\Rightarrow 25 + 36 - 49 = 2 \times 6$
$\Rightarrow 12 = 12$
Hence, the result is true.

> **HINTS AND TIPS**
>
> In parts **b** and **c**, you should start with one side and work towards an expression that equals the other side, step by step.

b) Expand the LHS and RHS of the expression:
LHS $= n^2 + n^2 + 2n + 1 - (n^2 + 4n + 4) = n^2 - 2n - 3$
RHS $= (n - 3)(n + 1) = n^2 - 2n - 3$
Therefore, LHS = RHS
That is, both sides are algebraically the same.

c) Expand the LHS of the expression: $\quad n^2 + n^2 + 2n + 1 - (n^2 + 4n + 4)$
Collect like terms: $\quad n^2 + n^2 - n^2 + 2n - 4n + 1 - 4 = n^2 - 2n - 3$
Factorise the collected result: $\quad = (n - 3)(n + 1)$
which is the RHS of the original expression.

EXAMPLE 2

Prove by deduction that the sum of the squares of three consecutive odd numbers is always one less than a multiple of 12.

First, write three consecutive odd numbers algebraically as $2n + 1$ and $2n + 3$ and $2n + 5$.
Write an expression for the sum of the squares: $(2n + 1)^2 + (2n + 3)^2 + (2n + 5)^2$
Square and add these expressions: $= 4n^2 + 4n + 1 + 4n^2 + 12n + 9 + 4n^2 + 20n + 25$
Simplify: $\quad = 12n^2 + 36n + 35$
Write 35 as $36 - 1$ as 36 is a
multiple of 12: $\quad = 12n^2 + 36n + 36 - 1$
Factorise, taking 12 outside
the bracket: $\quad = 12(n^2 + 3n + 3) - 1$
Leave the -1 outside the bracket to prove that the result is one less than a multiple of 12.

EXAMPLE 3

Prove that $\dfrac{\sqrt{3} - 2}{\sqrt{48} + 6} \equiv 2 - \dfrac{7\sqrt{3}}{6}$

For this question you need to rationalise the denominator of the left-hand side (LHS) of the identity and then simplify it to equal the right-hand side (RHS).

Simplify $\sqrt{48}$: $\qquad\qquad\qquad\qquad \dfrac{\sqrt{3}-2}{\sqrt{48}+6} = \dfrac{\sqrt{3}-2}{4\sqrt{3}+6}$

Rationalise the denominator $\dfrac{\sqrt{3}-2}{4\sqrt{3}+6} \times \dfrac{4\sqrt{3}-6}{4\sqrt{3}-6} = \dfrac{12 - 6\sqrt{3} - 8\sqrt{3} + 12}{48 - 36}$

Simplify: $\qquad\qquad\qquad\qquad\qquad\qquad\quad = \dfrac{24 - 14\sqrt{3}}{12}$

Write in the form required (as the RHS): $\quad = \dfrac{24}{12} - \dfrac{14\sqrt{3}}{12}$

$\qquad\qquad\qquad\qquad\qquad\qquad\qquad\qquad = 2 - \dfrac{7\sqrt{3}}{6}$

$\qquad\qquad\qquad\qquad\qquad\qquad\qquad\qquad = \text{RHS}$

Therefore, $\dfrac{\sqrt{3}-2}{\sqrt{48}+6} \equiv 2 - \dfrac{7\sqrt{3}}{6}$

HINTS AND TIPS

The symbol $\equiv$ represents an **identity**, which means that the left-hand side (LHS) is identical to the right-hand side (RHS) for all values of the variables. This is different from an equation, which is true for only some values of the variables.

EXERCISE 9A

TARGETING DISTINCTION

1. **a)** Choose any odd number and any even number. Add these together. Is the result odd or even? Does this always work for any odd number and even number that you choose?

   b) Let any odd number be represented by $2n + 1$. Let any even number be represented by $2m$, where m and n are integers. Prove that the sum of an odd number and an even number is always an odd number.

2. Prove the following results.

   a) The sum of two even numbers is even.

   b) The product of two even numbers is even.

   c) The product of an odd number and an even number is even.

   d) The product of two odd numbers is odd.

   e) The sum of four consecutive numbers is always even.

   f) Half the sum of four consecutive numbers is always odd.

3. A Fibonacci-type sequence is formed by adding the previous two terms to get the next term. For example, starting with 3 and 4, the series is:

   3, 4, 7, 11, 18, 29, 47, 76, 123, 199, ...

   a) Continue the Fibonacci sequence 1, 1, 2, ... up to 10 terms.

   b) Continue the Fibonacci-type sequence $a, b, a+b, a+2b, 2a+3b, \ldots$ up to 10 terms.

   c) Prove that the difference between the 8th term and the 5th term of any Fibonacci-type sequence is twice the 6th term.

4. The *n*th term in the sequence of triangular numbers 1, 3, 6, 10, 15, 21, 28, ... is given by $\frac{1}{2}n(n+1)$.

  a) Show that the sum of the 11th and 12th terms is a perfect square.

  b) Explain why the $(n+1)$th term of the triangular number sequence is given by $\frac{1}{2}(n+1)(n+2)$.

  c) Prove that the sum of any two consecutive triangular numbers is always a square number.

5. The sum of the series $1 + 2 + 3 + 4 + ... + (n-2) + (n-1) + n$ is given by $\frac{1}{2}n(n+1)$.

  a) Verify that this result is true for $n = 6$.

  b) Write down a simplified value, in terms of n, for the sum of these two series.

  $1 + 2 + 3 + ... + (n-2) + (n-1) + n$ and $n + (n-1) + (n-2) + ... + 3 + 2 + 1$

  c) Prove that the sum of the first n integers is $\frac{1}{2}n(n+1)$.

6. *T* represents any triangular number. Prove that:

  a) $8T + 1$ is always a square number

  b) $9T + 1$ is always another triangular number.

7. Lewis Carroll, who wrote *Alice in Wonderland*, was also a mathematician. In 1890, he suggested the following results.

  a) For any pair of numbers, x and y, if $x^2 + y^2$ is even, then $\frac{1}{2}(x^2 + y^2)$ is the sum of two squares.

  b) For any pair of numbers, x and y, $2(x^2 + y^2)$ is always the sum of two squares.

  c) Any number of which the square is the sum of two squares is itself the sum of two squares.

  Can you prove these statements to be true or false?

8. Pythagoras' theorem says that $a^2 + b^2 = c^2$ for a right-angled triangle with two short sides a and b and a long side c. For any integer n, the expressions $2n$, $n^2 - 1$ and $n^2 + 1$ form three numbers that obey Pythagoras' theorem. Can you prove this?

9. Waring's theorem states that: 'Any whole number can be written as the sum of not more than four square numbers.'

  For example, $27 = 3^2 + 3^2 + 3^2$ and $23 = 3^2 + 3^2 + 2^2 + 1^2$. Is this always true?

10. The difference of two squares is an identity: $a^2 - b^2 \equiv (a+b)(a-b)$, which means that it is true for all values of a and b, whether they are numerical or algebraic. Prove that $a^2 - b^2 \equiv (a+b)(a-b)$ is true when $a = 2x + 1$ and $b = x - 1$.

11. Prove that $\dfrac{\sqrt{12} + 5}{\sqrt{3} + 2} \equiv 4 - \sqrt{3}$

12. Prove that $\dfrac{1}{\frac{1}{\sqrt{5}} + \sqrt{5}} \equiv \dfrac{\sqrt{5}}{6}$

EXAMPLE 4

Prove that the function $y = 3x^2 - 4x + 7$ is positive for all values of x.

$$3x^2 - 4x + 7 = 3\left(x^2 - \frac{4}{3}\right) + 7$$
$$= 3\left(\left(x - \frac{2}{3}\right)^2 - \frac{4}{9}\right) + 7$$
$$= 3\left(x - \frac{2}{3}\right)^2 - \frac{4}{3} + 7$$
$$= 3\left(x - \frac{2}{3}\right)^2 + \frac{17}{3}$$

> **HINTS AND TIPS**
>
> Remember that a quadratic function with a positive x squared term (i.e. $a \geqslant 0$) will have a minimum point as it has a 'U' shape.

This must be positive as any squared term is positive.

The minimum (or turning) point of the function is $\left(\frac{2}{3}, \frac{17}{3}\right)$ which is greater than 0.

Therefore, the function $y = 3x^2 - 4x + 7$ is positive for all values of x (the range is $y \geqslant \frac{17}{3}$).

EXERCISE 9B

TARGETING DISTINCTION*

1. Prove that the function $y = \frac{x^2}{3} - 2x + 5$ is positive for all values of x.
2. Prove that the function $y = -2x^2 - x - 3$ is negative for all values of x.

> **HINTS AND TIPS**
>
> Remember that a quadratic function with a negative x squared term (i.e. $a \leqslant 0$) will have a maximum point as it has an '∩' shape.

9.2 Proof by exhaustion and disproof by counter example

THIS SECTION WILL SHOW YOU HOW TO ...
✓ understand how to prove by exhaustion
✓ understand how to disprove by counter example

KEY WORDS
✓ prove by exhaustion
✓ cases
✓ disprove by counter example

To **prove by exhaustion**, you look at individual **cases** or occurrences and show that they are all true and agree with the statement.

This must cover all the possible situations defined in the question.

EXAMPLE 5

Prove by exhaustion that $n^2 + n + 17$ is prime for all integers from 1 to 6.

To prove by exhaustion, you need to check all the cases (or possibilities). In this example, this means every integer from 1 to 6.

$n = 1$	$1^2 + 1 + 17 = 19$	19 is prime	$n = 4$	$4^2 + 4 + 17 = 37$	37 is prime
$n = 2$	$2^2 + 2 + 17 = 23$	23 is prime	$n = 5$	$5^2 + 5 + 17 = 47$	47 is prime
$n = 3$	$3^2 + 3 + 17 = 29$	29 is prime	$n = 6$	$6^2 + 6 + 17 = 59$	59 is prime

As every integer from 1 to 6 results in a prime number, you can say that the statement has been proved by exhaustion.

EXAMPLE 6

Prove by exhaustion that $n^2 + 1$ is not divisible by 4 for $0 < n \leq 7$.

$n = 1$	$1^2 + 1 = 2$	not divisible by 4	$n = 5$	$5^2 + 1 = 26$	not divisible by 4
$n = 2$	$2^2 + 1 = 5$	not divisible by 4	$n = 6$	$6^2 + 1 = 37$	not divisible by 4
$n = 3$	$3^2 + 1 = 10$	not divisible by 4	$n = 7$	$7^2 + 1 = 50$	not divisible by 4
$n = 4$	$4^2 + 1 = 17$	not divisible by 4			

Each result is not divisible by 4, so the statement is proved by exhaustion.

To **prove by counter example**, you need to find an example that results in the statement being false.

Usually, a numerical example will be the easiest way to answer these types of questions. You may need to try several different numbers until you find one that disproves the statement.

EXAMPLE 7

Prove by counter example that $n^2 + n + 5$ is not prime for all values of n.

$n = 5$ $5^2 + 5 + 5 = 35$ 35 is not prime

This is enough for the proof, as a counter example has been found.

EXERCISE 9C

TARGETING DISTINCTION

1. Prove by exhaustion that the product of two consecutive odd numbers is odd for all integers from 1 to 7.
2. Prove by exhaustion that 17 is a prime number.

3. Prove by exhaustion that there are six distinct ways of arranging the letters TOOT.

4. Disprove by counter example that the square root of the square of x equals x.

5. Disprove by counter example that the difference between any two square numbers is always odd.

EXAMPLE 8

Prove by exhaustion that $x^3 + x + 3$ is an odd integer.

For a general proof like this you can split the question into testing odd numbers and testing even numbers algebraically.

If x is even, then substitute $2n$: $\quad (2n)^3 + 2n + 3 \Rightarrow 8n^3 + 2n + 3$

Now factorise: $\quad \Rightarrow 2(4n^3 + n + 1) + 1$

(always odd)

If x is odd, then substitute $2m + 1$: $(2m + 1)^3 + (2m + 1) + 3 \Rightarrow 8m^3 + 12m^2 + 6m + 1 + 2m + 1 + 3$

Simplify: $\quad \Rightarrow 8m^3 + 12m^2 + 8m + 5$

Now factorise: $\quad \Rightarrow 2(4m^3 + 6m^2 + 4m + 2) + 1$

The result is always odd as it in in the form $2n + 1$.

This proves that when x is even $x^3 + x + 3$ is an odd integer and when x is odd $x^3 + x + 3$ is an odd integer.

All integers must be either odd or even, so the statement is proved for all possible cases.

HINTS AND TIPS

When proving separately for even and odd numbers you can use two different letters, for example, n and m to differentiate between them.

EXERCISE 9D

TARGETING DISTINCTION*

1. Prove by exhaustion that $x^3 + 3x + 1$ is an odd integer.
2. Prove by exhaustion that $2^n - 1$ is always prime for $1 < n < 4$.
3. Prove by counter example that $2^n - 1$ is not always prime.

HINTS AND TIPS

Remember that a counter example is an exception to a rule. Try lower (or familiar) values first: for example, start at $n = 1$ and work up.

4. Prove that for all positive integers n such that $1 \leq n \leq 4$, the expression $n^2 + n$ is divisible by 2.
5. Prove by counter example that $\sin 2\theta$ does not always equal $2\sin\theta$.
6. Prove by exhaustion that the differences of the squares of any two consecutive even numbers less than 10 is divisible by 4.

9.3 Geometric proofs

THIS SECTION WILL SHOW YOU HOW TO ...
✓ understand geometric proofs

KEY WORDS
✓ geometric

You should already know:
- the angle sum of the interior angles in a triangle (180°)
- the circle theorems
- Pythagoras' theorem.

Can you prove these results?

For a **geometric** proof, you must proceed in logical steps, to establish a series of mathematical statements by using facts that are already known to be true.

Proof that the sum of the interior angles of a triangle is 180°

Look at the following proof.

Start with triangle ABC with angles α, β and γ (Figure 1).

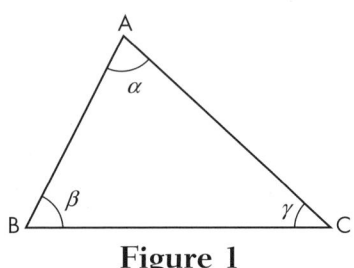

Figure 1

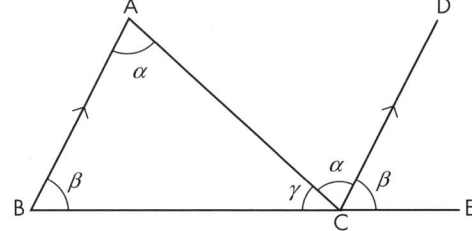

Figure 2

On Figure 1, draw a line CD parallel to side AB and extend BC to E, to give Figure 2.

Since AB is parallel to CD:

$\angle ACD = \angle BAC = \alpha$ (alternate angles)

$\angle DCE = \angle ABC = \beta$ (corresponding angles)

BCE is a straight line, so $\gamma + \alpha + \beta = 180°$. Therefore, the sum of the interior angles of a triangle is 180°.

This proof assumes that alternate angles are equal and that corresponding angles are equal. Strictly, you should prove these results, but you have to accept certain results as true. These are based on Euclid's axioms from which all geometric proofs are derived.

Proof of Pythagoras' theorem

Draw a square of side c inside a square of side $(a + b)$, as shown.

The area of the exterior square is $(a + b)^2 = a^2 + 2ab + b^2$.

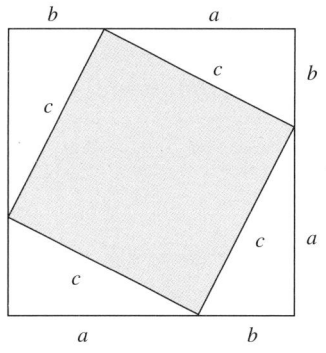

The area of each small triangle around the shaded square is $\frac{1}{2}ab$.

The total area of all four triangles is $4 \times \frac{1}{2}ab = 2ab$.

Subtracting the total area of the four triangles from the area of the large square gives the area of the shaded square:

$a^2 + 2ab + b^2 - 2ab = a^2 + b^2$

But the area of the shaded square is c^2, so $c^2 = a^2 + b^2$, which is Pythagoras' theorem.

Congruence proof

There are four conditions to prove congruence. These are known as SSS (three sides the same), SAS (two sides and the included angle the same), ASA (or AAS) (two angles and one corresponding side the same) and RHS (right-angled triangle, hypotenuse and one short side the same).

Note: AAA (three angles the same) is *not* a condition for congruence.

When you prove a result, you must explain or justify every statement or line. Proofs have to be rigorous and logical.

EXAMPLE 9

ABCD is a parallelogram. X is the point where the diagonals meet.

Prove that triangles AXB and CXD are congruent.

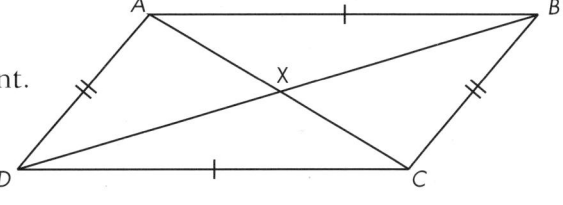

∠BAX = ∠DCX (alternate angles)
∠ABX = ∠CDX (alternate angles)
AB = CD (opposite sides in a parallelogram)
Hence △AXB is congruent to △CXD (ASA).

Note: You could have used ∠AXB = ∠CXD (vertically opposite angles) as the second line. Whichever approach is used you must give a reason for each statement.

Circle theorems.

You need to be able to prove circle theorems.

Here is a reminder of the some of the standard circle theorems.

The angle at the centre of a circle is twice the angle at the circumference.

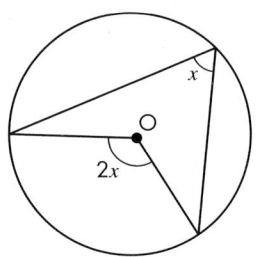

The angle subtended by a diameter at the circumference of a circle is a right angle.

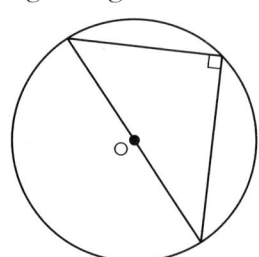

Opposite angles in a cyclic quadrilateral add to 180°.

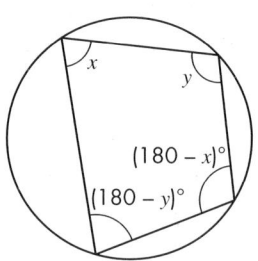

Angles subtended by the same chord (or arc) are equal.

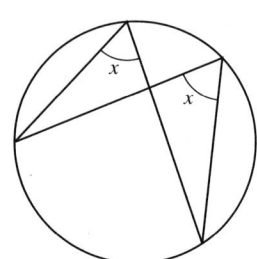

The angle between a radius and a tangent is 90°.

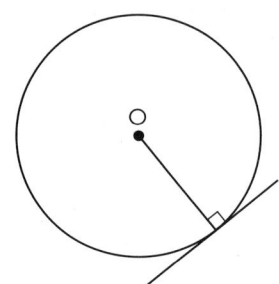

The alternate segment theorem

The angle formed between a tangent and a chord is equal to the angle in the alternate segment (subtended by the chord at the circumference).

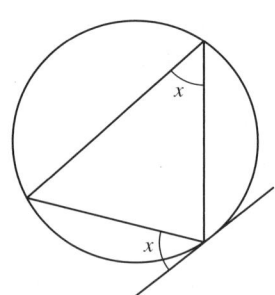

EXAMPLE 10

D, E and F are points on the circumference of a circle centre C.

Prove that angle ECF is twice the size of angle EDF.

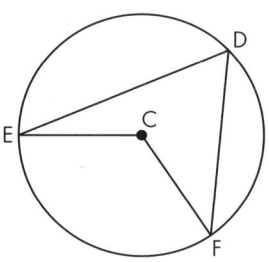

EC, CF and DC are all equal length as they are radii.
So, CED and CDF are isosceles triangles
$\Rightarrow$ angle DEC = angle CDE = x
and angle CDF = angle CFD = y
$\Rightarrow$ angle DCE = $180° - 2x$ and angle DCF = $180° - 2y$
$\Rightarrow 360°$ = angle ECF + angle DCE + angle DCF
$\Rightarrow 360°$ = angle ECF + $180° - 2x + 180° - 2y$
$\Rightarrow 360°$ = angle ECF + $360° - 2x - 2y$

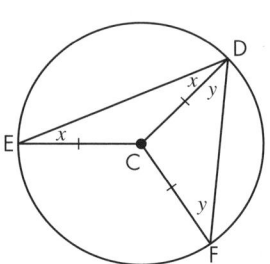

⇒ $2x + 2y$ = angle ECF
⇒ $2(x + y)$ = angle ECF
As $x + y$ = angle EDF ⇒ 2 × angle EDF = angle ECF
This is proved, as required.

HINTS AND TIPS

When proving circle theorems, draw radii (and/or a diameter) to formulate the proof.
Remember that radii form isosceles triangles with equal base angles.

EXAMPLE 11

Prove that the opposite angles in a cyclic quadrilateral sum to 180°.

First, draw a cyclic quadrilateral and then draw two radii from the centre to the circumference of the circle as shown.

From the previous proof you know that the angle at the centre of a circle is twice the angle at the circumference, so you can label the angles shown on the diagram.

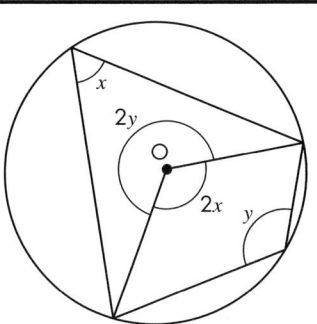

$2x + 2y = 360°$ (from the centre of the circle)
$2(x + y) = 360°$ ⇒ $x + y = 180°$

This proves that the opposite angles in the cyclic quadrilateral sum to 180°.

EXERCISE 9E

TARGETING DISTINCTION

1. Prove that the triangle DEF with one angle of $x°$ and an exterior angle of $90° + \frac{x°}{2}$ is isosceles.

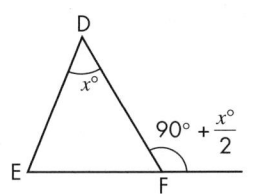

2. Prove that a triangle with an interior angle of $\frac{x°}{2}$ and an exterior angle of $x°$ is isosceles.

3. A, B and C are points on the circumference of a circle, centre O. AOC is a diameter of the circle.

   Prove that angle ABC is 90°

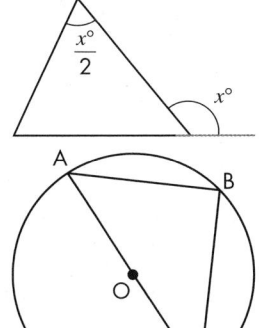

4. Prove that angle ACB = angle CED.

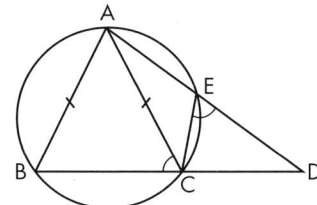

5. PQRS is a parallelogram. Prove that triangles PQS and RQS are congruent.

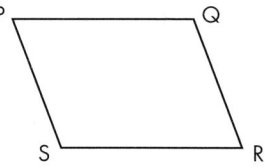

6. Prove by counter example that not all equilateral triangles are obtuse.

TARGETING DISTINCTION*

7. a) Prove the alternate segment theorem.
   b) Two circles touch internally at T. The common tangent at T is drawn. Two lines TAB and TXY are drawn from T. Prove that AX is parallel to BY.

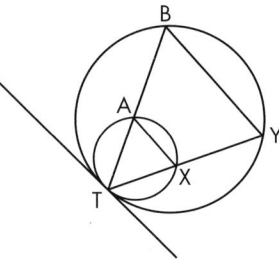

Exam-style questions

1. You are given that: $(a + b)^2 + (a - b)^2 = 2(a^2 + b^2)$.
   a) Verify that this result is true for $a = 3$ and $b = 4$. [2 marks]
   b) Show that the LHS is the same as the RHS. [2 marks]
   c) Prove that the LHS can be simplified to the RHS. [2 marks]

2. Prove that: $(a + b)^2 - (a - b)^2 = 4ab$. [3 marks]

3. 'When two numbers have a difference of 2, the difference of their squares is twice the sum of the two numbers.'
   a) Verify that this is true for 5 and 7. [2 marks]
   b) Prove that the result is true. [4 marks]
   c) Prove that when two numbers have a difference of n, the difference of their squares is n times the sum of the two numbers. [3 marks]

4. Four consecutive numbers are 4, 5, 6 and 7.
   a) Verify that their product plus 1 is a perfect square. [2 marks]
   b) Use a suitable method to show that:
   $(n^2 - n - 1)^2 = n^4 - 2n^3 - n^2 + 2n + 1$. [3 marks]
   c) Let four consecutive numbers be $(n - 2), (n - 1), n, (n + 1)$. Prove that the product of four consecutive numbers plus 1 is a perfect square. [4 marks]

5. Prove that the sum of the squares of two consecutive integers is an odd number. [4 marks]

6. The square of the sum of the first n consecutive whole numbers is equal to the sum of the cubes of the first n consecutive whole numbers.
  a) Verify that $(1 + 2 + 3 + 4)^2 = 1^3 + 2^3 + 3^3 + 4^3$. [2 marks]
  b) The sum of the first n consecutive whole numbers is $\frac{1}{2}n(n + 1)$. Write down a formula for the sum of the cubes of the first n whole numbers. [1 mark]
  c) Test your formula for $n = 6$. [2 marks]

7. Two circles touch externally at T. Line ATB is drawn through T. The common tangent at T and the tangents at A and B meet at P and Q. Prove that PB is parallel to AQ. [4 marks]

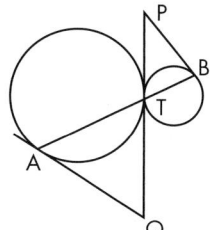

8. Prove that $n^3 - n$ is equal to the product of three consecutive numbers. [3 marks]

9. Prove that the angle subtended by the diameter of a circle is $90°$. [3 marks]

10. The diagram on the right shows a circle with centre O and points A, B, C, D on the circumference.

Prove that angle ABD equals angle ACD. [2 marks]

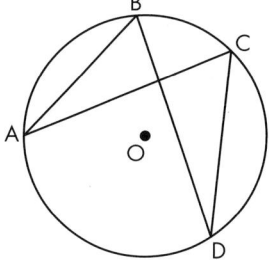

10 Vectors

10.1 Position vectors

THIS SECTION WILL SHOW YOU HOW TO …
- ✓ add and subtract vectors
- ✓ multiply vectors by a scalar
- ✓ find the distance between position vectors
- ✓ understand that parallel vectors are multiples of each other

KEY WORDS
- ✓ vector
- ✓ resultant vector
- ✓ position vector
- ✓ parallel
- ✓ scalar
- ✓ collinear

A **vector** quantity has both a magnitude (a number that represents its size) and a direction. You can visualise a vector as a directed line segment. The length of the segment corresponds to the magnitude of the vector, and an arrow indicates its direction. For example, a vector pointing diagonally to the right with a length of a units can be represented graphically by a line with an arrowhead pointing right, scaled to a length of a units.

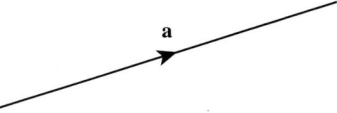

Usually, vectors are labelled with lower case letters in bold. When you are writing vectors you should underline your lowercase vector labels so that they are clear (as it is difficult to apply 'bold' when handwriting).

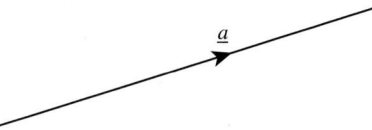

Because they can represent distances and directions, vectors have many everyday applications, for example, in global positioning systems (GPS). Vectors have links to molecular biology (as a means of carrying a particular DNA segment into a host cell) and they are used in physics to represent displacement, acceleration, force, velocity and momentum. Vectors are also used in the storage of locations, directions and velocities in video games and in engineering to analyse mechanical systems.

Scalar quantities have only magnitude (or size), for example, any number, a temperature or a mass.

HINTS AND TIPS

You met vectors when you translated objects on a coordinate grid. Recall that the vector $\begin{pmatrix} 4 \\ 5 \end{pmatrix}$ indicates a horizontal movement of 4 units to the right and a vertical movement of 5 units upwards. The numbers 4 and 5 are the horizontal and vertical components of the vector.

Operations on vectors

Vectors can undergo a variety of operations, each of which has practical applications.

Addition and subtraction

Vectors are added or subtracted by adding or subtracting their corresponding components. The sum of two vectors is called the **resultant vector**.

If $\mathbf{a} = \begin{pmatrix} x_1 \\ y_1 \end{pmatrix}$ and $\mathbf{b} = \begin{pmatrix} x_2 \\ y_2 \end{pmatrix}$, then their sum $\mathbf{c} = \mathbf{a} + \mathbf{b}$ is given by

$\mathbf{c} = \begin{pmatrix} x_1 + x_2 \\ y_1 + y_2 \end{pmatrix}$ and their difference $\mathbf{d} = \mathbf{a} - \mathbf{b}$ as $\mathbf{d} = \begin{pmatrix} x_1 - x_2 \\ y_1 - y_2 \end{pmatrix}$.

These operations can be visualised geometrically. Vectors are added by drawing 'nose-to-tail', as shown below.

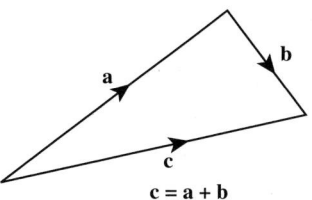

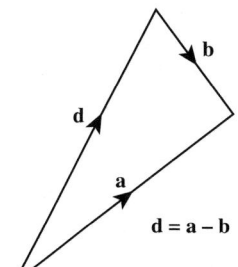

The arrows on $\mathbf{a}$ and $\mathbf{b}$ form a continuous path. The arrow on the resultant $\mathbf{c}$ shows that the vector goes from the start of $\mathbf{a}$ to the end of $\mathbf{b}$.

A negative vector has the same length but in the opposite direction. So to subtract a vector you move along it in the opposite direction to the arrow.

When you add two vectors the order does not matter, so $\mathbf{a} + \mathbf{b} = \mathbf{b} + \mathbf{a}$.

Scalar multiplication

A vector can be multiplied by a scalar (a real number) by multiplying each of its components by that scalar.

If $\mathbf{e} = \begin{pmatrix} x \\ y \end{pmatrix}$ and k is a scalar, then the vector $k\mathbf{e} = k\begin{pmatrix} x \\ y \end{pmatrix} = \begin{pmatrix} kx \\ ky \end{pmatrix}$.

This operation changes the magnitude of the vector but not its direction. However, if the scalar is negative, then the direction of the vector is reversed.

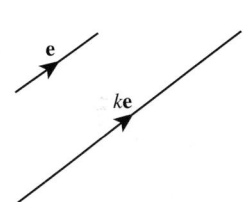

EXAMPLE 1

$\mathbf{a} = \begin{pmatrix} -2 \\ 7 \end{pmatrix}$ and $\mathbf{b} = \begin{pmatrix} 3 \\ 5 \end{pmatrix}$

Work out $3\mathbf{a} - 2\mathbf{b}$.

$$3\mathbf{a} - 2\mathbf{b} = 3\begin{pmatrix} -2 \\ 7 \end{pmatrix} - 2\begin{pmatrix} 3 \\ 5 \end{pmatrix}$$

Multiply each vector by the scalar:

$$= \begin{pmatrix} 3 \times -2 \\ 3 \times 7 \end{pmatrix} - \begin{pmatrix} 2 \times 3 \\ 2 \times 5 \end{pmatrix}$$

Then do the calculation:

$$= \begin{pmatrix} -6 \\ 21 \end{pmatrix} - \begin{pmatrix} 6 \\ 10 \end{pmatrix}$$

$$= \begin{pmatrix} -6 - 6 \\ 21 - 10 \end{pmatrix}$$

$$= \begin{pmatrix} -12 \\ 11 \end{pmatrix}$$

HINTS AND TIPS

Combine the two brackets to give a single vector. Be careful when calculating with negative numbers.

EXAMPLE 2

$\mathbf{c} = \begin{pmatrix} 4 \\ -2 \end{pmatrix}$ and $\mathbf{d} = \begin{pmatrix} 3 \\ 3 \end{pmatrix}$

Given that $3\mathbf{c} - 2\mathbf{e} = -2\mathbf{d}$, find $\mathbf{e}$.

Substitute the values into the equation:
$$3\begin{pmatrix} 4 \\ -2 \end{pmatrix} - 2\mathbf{e} = -2\begin{pmatrix} 3 \\ 3 \end{pmatrix}$$

$$\begin{pmatrix} 12 \\ -6 \end{pmatrix} - 2\mathbf{e} = \begin{pmatrix} -6 \\ -6 \end{pmatrix}$$

Rearrange and solve:
$$\begin{pmatrix} 12 \\ -6 \end{pmatrix} - \begin{pmatrix} -6 \\ -6 \end{pmatrix} = 2\mathbf{e}$$

$$\begin{pmatrix} 18 \\ 0 \end{pmatrix} = 2\mathbf{e}$$

$$\begin{pmatrix} 9 \\ 0 \end{pmatrix} = \mathbf{e}$$

Finding the distance between two points given by position vectors

Vectors can represent the position of an object relative to a fixed point, for example, the origin on a coordinate grid. You can find the distance between two **position vectors** using Pythagoras' theorem. This is the same method as you use to find the distance between two points whose coordinates you know.

This gives the formula

$d = \sqrt{(x_2 - x_1)^2 + (y_2 - y_1)^2}$, where (x_1, y_1) and (x_2, y_2) are the coordinates of two points and d is the distance between them.

EXAMPLE 3

Find the exact distance between the points given by position vectors

$\mathbf{a} = \begin{pmatrix} 4 \\ 2 \end{pmatrix} = \begin{pmatrix} x_1 \\ y_1 \end{pmatrix}$ and $\mathbf{b} = \begin{pmatrix} -1 \\ 6 \end{pmatrix} = \begin{pmatrix} x_2 \\ y_2 \end{pmatrix}$.

You can represent these vectors on a coordinate grid and draw a right-angled triangle to help you to calculate the distance d between them.

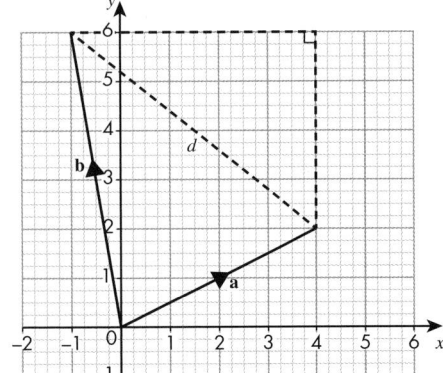

$$\mathbf{a} = \begin{pmatrix} 4 \\ 2 \end{pmatrix} = \begin{pmatrix} x_1 \\ y_1 \end{pmatrix} \text{ and } \mathbf{b} = \begin{pmatrix} -1 \\ 6 \end{pmatrix} = \begin{pmatrix} x_2 \\ y_2 \end{pmatrix}$$

Substitute the numbers into the formula: $d = \sqrt{(x_2 - x_1)^2 + (y_2 - y_1)^2}$

This gives
$$d = \sqrt{(-1 - 4)^2 + (6 - 2)^2}$$
$$= \sqrt{(-5)^2 + (4)^2}$$
$$= \sqrt{25 + 16}$$
$$= \sqrt{41} \text{ (take the positive square root only)}$$

The diagram shows the calculation geometrically, but it is not necessary for you to draw a diagram if you use the formula.

The question asks for the **exact** distance, so leave your answer in surd form (as rounding to a number of decimal places would lose accuracy and so would not be an exact answer).

EXERCISE 10A

TARGETING PASS

1. $\mathbf{a} = \begin{pmatrix} 3 \\ -2 \end{pmatrix}$ $\mathbf{b} = \begin{pmatrix} -3 \\ 4 \end{pmatrix}$ $\mathbf{c} = \begin{pmatrix} 0 \\ 7 \end{pmatrix}$ $\mathbf{d} = \begin{pmatrix} -1 \\ -1 \end{pmatrix}$

   Work out:
   a) $\mathbf{a} + \mathbf{b}$ b) $\mathbf{c} - \mathbf{d}$ c) $\mathbf{a} - \mathbf{b} + \mathbf{c}$
   d) $2\mathbf{a} + 3\mathbf{c}$ e) $-4\mathbf{d} - 2\mathbf{c}$ f) $3\mathbf{c} + 4\mathbf{a} - 2\mathbf{b}$

2. WXYZ is a trapezium, $\overrightarrow{WX} = \mathbf{a}$, $\overrightarrow{ZY} = 3\mathbf{a}$, $\overrightarrow{WZ} = \mathbf{b}$

   V is the midpoint of XY.
   Write down each vector, in terms of **a** and **b**.
   a) $\overrightarrow{XZ}$ b) $\overrightarrow{XY}$ c) $\overrightarrow{YX}$ d) $\overrightarrow{VX}$ e) $\overrightarrow{ZV}$

   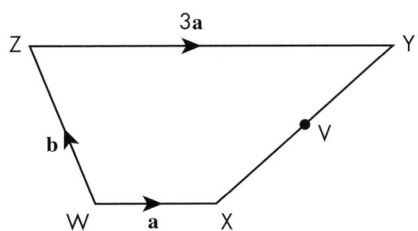

3. $\mathbf{e} = \begin{pmatrix} 3 \\ -2 \end{pmatrix}, \mathbf{f} = \begin{pmatrix} -3 \\ x \end{pmatrix}$

   Solve the equation to find x.

   $3\mathbf{e} - 2\mathbf{f} = 3\begin{pmatrix} 5 \\ 1 \end{pmatrix}$

4. $\mathbf{g} = \begin{pmatrix} x \\ y \end{pmatrix}, \mathbf{h} = \begin{pmatrix} y \\ x \end{pmatrix}$ and $2\mathbf{g} + 3\mathbf{h} = \begin{pmatrix} 7 \\ 3 \end{pmatrix}$

   Form a pair of simultaneous equations and work out the value of x and the value of y.

5. $j = \begin{pmatrix} 3 \\ -2 \end{pmatrix}$, $k = \begin{pmatrix} -3 \\ 4 \end{pmatrix}$

Given that $\mathbf{m} - 3\mathbf{j} = 2\mathbf{k}$, work out $\mathbf{m}$.

6. Find the exact distance between the points given by position vectors $\mathbf{a} = \begin{pmatrix} -2 \\ 3 \end{pmatrix}$ and $\mathbf{b} = \begin{pmatrix} 5 \\ 2 \end{pmatrix}$.

7. Find the distance between the position vectors $\mathbf{a} = \begin{pmatrix} 6 \\ -5 \end{pmatrix}$ and $\mathbf{b} = \begin{pmatrix} 2 \\ 7 \end{pmatrix}$ giving your answer to 2 dp.

> **HINTS AND TIPS**
>
> A vector from point A to point B is denoted by $\overrightarrow{AB}$. You should use this notation in your working.

Parallel vectors

Parallel vectors have the same direction, but may have different magnitudes. Vectors are parallel if they are scalar multiples of each other.

- If you multiply a vector by a real positive number, then the result is a vector with a different magnitude but the same direction
- If you multiply a vector by a real negative number, then the result is a vector with a different magnitude but the opposite direction.

This means that parallel vectors have a common factor. For example:

- $\mathbf{a} = \begin{pmatrix} 6 \\ -3 \end{pmatrix}$ and $\mathbf{b} = \begin{pmatrix} 10 \\ -5 \end{pmatrix}$ are parallel because they can be written as $\mathbf{a} = 3\begin{pmatrix} 2 \\ -1 \end{pmatrix}$ and $\mathbf{b} = 5\begin{pmatrix} 2 \\ -1 \end{pmatrix}$. So they are multiples of the same vector $\begin{pmatrix} 2 \\ -1 \end{pmatrix}$.

- $4\mathbf{a} + 6\mathbf{b}$ and $6\mathbf{a} + 9\mathbf{b}$ can be written as $2(2\mathbf{a} + 3\mathbf{b})$ and $3(2\mathbf{a} + 3\mathbf{b})$.

  They share the factor $(2\mathbf{a} + 3\mathbf{b})$ and so are parallel.

EXAMPLE 4

Show that the position vectors $\mathbf{a} = \begin{pmatrix} 4 \\ -2 \end{pmatrix}$ and $\mathbf{b} = \begin{pmatrix} -6 \\ 3 \end{pmatrix}$ are parallel.

$\mathbf{a} = 2\begin{pmatrix} 2 \\ -1 \end{pmatrix}$ and $\mathbf{b} = -3\begin{pmatrix} 2 \\ -1 \end{pmatrix}$ (The ratio of $\mathbf{a}$ to $\mathbf{b}$ is $2 : -3$.)

$\mathbf{a}$ and $\mathbf{b}$ are multiples of the same vector, $\begin{pmatrix} 2 \\ -1 \end{pmatrix}$ and $\mathbf{a} = -\frac{2}{3}\mathbf{b}$; therefore they are parallel.

Collinear points

Three points are **collinear** if they all lie on a single straight line.
If vectors $\overrightarrow{XY}$ and $\overrightarrow{YZ}$ are parallel, then the points X, Y and Z must be collinear.

EXAMPLE 5

$\overrightarrow{OX} = \begin{pmatrix} 3 \\ -2 \end{pmatrix}, \overrightarrow{OY} = \begin{pmatrix} 7 \\ 3 \end{pmatrix}, \overrightarrow{OZ} = \begin{pmatrix} 15 \\ 13 \end{pmatrix}$

Show that X, Y and Z are collinear.

It helps to sketch a diagram:

$\overrightarrow{XY} = -\overrightarrow{OX} + \overrightarrow{OY} = -\begin{pmatrix} 3 \\ -2 \end{pmatrix} + \begin{pmatrix} 7 \\ 3 \end{pmatrix} = \begin{pmatrix} 4 \\ 5 \end{pmatrix}$

$\overrightarrow{YZ} = -\overrightarrow{OY} + \overrightarrow{OZ} = -\begin{pmatrix} 7 \\ 3 \end{pmatrix} + \begin{pmatrix} 15 \\ 13 \end{pmatrix} = \begin{pmatrix} 8 \\ 10 \end{pmatrix} = 2\begin{pmatrix} 4 \\ 5 \end{pmatrix}$

$2\overrightarrow{XY} = \overrightarrow{YZ}$, and so the vectors are parallel ($\overrightarrow{YZ}$ is a scalar multiple of $\overrightarrow{XY}$).

$\overrightarrow{XY}$ and $\overrightarrow{YZ}$ also share the point Y, and so X, Y and Z must be collinear (lie on the same straight line).

EXERCISE 10B

TARGETING MERIT

1. $\mathbf{a} = \begin{pmatrix} -6 \\ -2 \end{pmatrix}$ and $\mathbf{b} = \begin{pmatrix} 15 \\ 5 \end{pmatrix}$

   a) Show that the position vectors **a** and **b** are parallel.

   b) Write the ratio of **a** : **b**.

2. The points A, B and C have position vectors $\begin{pmatrix} 2 \\ -3 \end{pmatrix}, \begin{pmatrix} -1 \\ 5 \end{pmatrix}$ and $\begin{pmatrix} 4 \\ -2 \end{pmatrix}$, respectively.

   Find the position vector of D such that $\overrightarrow{AB} = \overrightarrow{CD}$.

3. $\overrightarrow{OA} = \begin{pmatrix} -5 \\ 1 \end{pmatrix}, \overrightarrow{OB} = \begin{pmatrix} -3 \\ 2 \end{pmatrix}, \overrightarrow{OC} = \begin{pmatrix} 3 \\ 5 \end{pmatrix}$

   Show that A, B and C are collinear.

4. $\overrightarrow{OD} = \mathbf{a} + \mathbf{b}, \overrightarrow{OE} = -2\mathbf{a} + 3\mathbf{b}, \overrightarrow{OF} = -11\mathbf{a} + 9\mathbf{b}$

   Show that D, E and F are collinear.

5. ABCD is a parallelogram. $\overrightarrow{AB} = -2\mathbf{a} + 2\mathbf{b}$, $\overrightarrow{BC} = 3\mathbf{a} + 4\mathbf{b}$.

   BDE is a straight line with $\overrightarrow{BD} = \overrightarrow{DE}$.
   Find $\overrightarrow{AB}$.

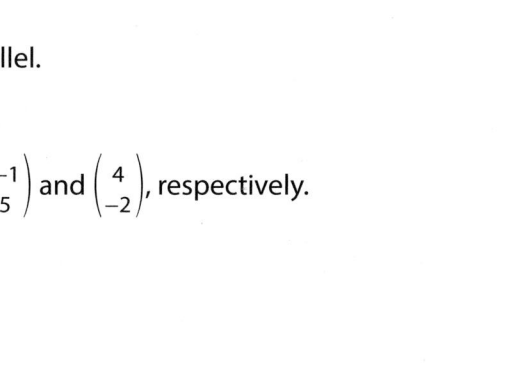

10.2 Solving geometric problems

THIS SECTION WILL SHOW YOU HOW TO ...
- ✓ solve geometric problems using vectors
- ✓ use vectors to construct geometric arguments and proofs

KEY WORDS
- ✓ geometric
- ✓ argument

Vectors can be used to solve **geometric** or spatial problems.

You may be asked to prove a vector problem by constructing an **argument**. This means a sequence of steps or statements (with reasons) that demonstrate that a claim is true (or false).

EXAMPLE 6

The diagram shows a triangle with an extended straight line OAE.

$OA = \frac{1}{2}AE$

$OB : OC = 2 : 1$

$\overrightarrow{OA} = \mathbf{a}$ and $\overrightarrow{OB} = \mathbf{b}$

$\overrightarrow{AD} = k\overrightarrow{AB}$ where k is a scalar quantity

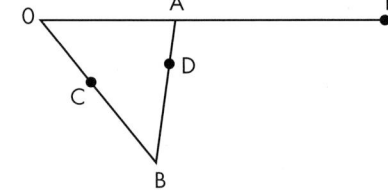

Construct a geometric argument to prove that if CDE is a straight line, then $k = \frac{2}{5}$.

The key to this argument is that $\overrightarrow{CD}$ is a multiple of $\overrightarrow{CE}$ (as they are collinear)

$$\overrightarrow{OE} = \overrightarrow{OA} + \overrightarrow{AE} = 3\mathbf{a} \text{ (as OA} = \tfrac{1}{2}\text{AE, so AE} = 2\text{OA)}$$

$$\overrightarrow{CE} = \overrightarrow{CO} + \overrightarrow{OE}$$

$$= -\tfrac{1}{2}\mathbf{b} + 3\mathbf{a}$$

$$= 3\mathbf{a} - \tfrac{1}{2}\mathbf{b}$$

$$\overrightarrow{CD} = \overrightarrow{CO} + \overrightarrow{OA} + \overrightarrow{AD}$$

$$= \overrightarrow{CO} + \overrightarrow{OA} + k\overrightarrow{AB} \text{ } (\overrightarrow{AB} = \overrightarrow{AO} + \overrightarrow{OB} = -\mathbf{a} + \mathbf{b} = \mathbf{b} - \mathbf{a})$$

$$= -\tfrac{1}{2}\mathbf{b} + \mathbf{a} + k\mathbf{b} - k\mathbf{a}$$

Simplify and factorise: $\quad = (1-k)\mathbf{a} + \left(k - \tfrac{1}{2}\right)\mathbf{b}$

You know that $\overrightarrow{CD}$ is a multiple of $\overrightarrow{CE}$, so the ratios of the coefficients of $\mathbf{a}$ and $\mathbf{b}$ will be equal.

This means that $\dfrac{1-k}{3} = \dfrac{k - \frac{1}{2}}{-\frac{1}{2}}$

Eliminate the fraction denominator on the RHS by dividing $k - \frac{1}{2}$ by $-\frac{1}{2}$ (which is the same as multiplying by -2). This gives

$$\dfrac{1-k}{3} = -2k + 1$$

Multiply both sides by 3: $\quad 1 - k = -6k + 3$

Solve to find k: $\quad\quad\quad\quad 5k = 2$

$$k = \tfrac{2}{5}$$

And the result is shown.

EXAMPLE 7

Find the angle between the vectors $\overrightarrow{AB} = \begin{pmatrix} 3 \\ 4 \end{pmatrix}$ and $\overrightarrow{AC} = \begin{pmatrix} 5 \\ -12 \end{pmatrix}$ correct to 1 decimal place.

You can use the cosine rule to find the angle between two vectors.

It is helpful to draw a diagram to visualise the problem.
First, calculate the lengths of vectors $\overrightarrow{AB}$ and $\overrightarrow{AC}$ using Pythagoras' theorem.

The length of $AB = \sqrt{3^2 + 4^2} = 5$

The length of $AC = \sqrt{5^2 + (-12)^2} = 13$

You can find the length of the vector $\overrightarrow{BC}$.

$\overrightarrow{BC} = \overrightarrow{AC} - \overrightarrow{AB} = \begin{pmatrix} 5 \\ -12 \end{pmatrix} - \begin{pmatrix} 3 \\ 4 \end{pmatrix} = \begin{pmatrix} 2 \\ -16 \end{pmatrix}$

You can calculate the length of $\overrightarrow{BC}$ (as before) using Pythagoras' theorem:

$\Rightarrow \sqrt{2^2 + (-16)^2} = \sqrt{260} = 2\sqrt{65}$

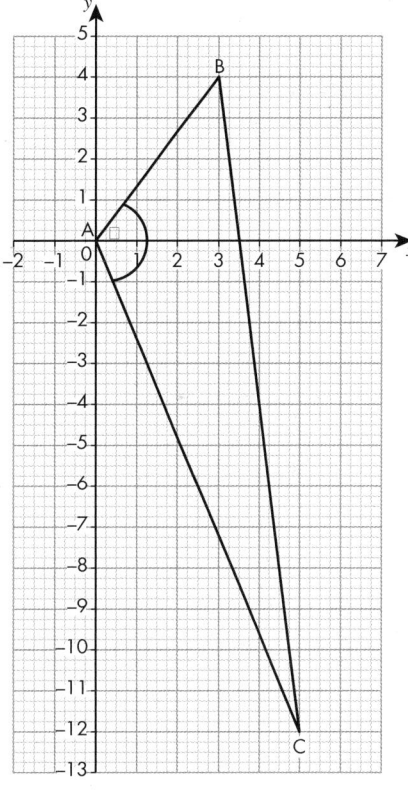

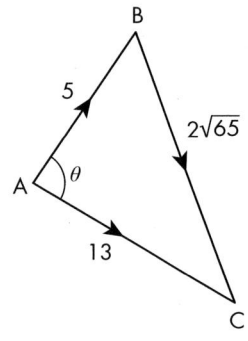
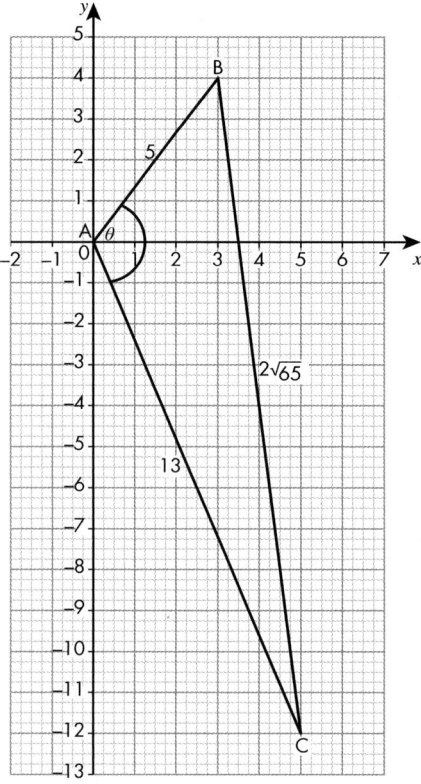

You can use the cosine rule to find the angle between $\vec{AB}$ and $\vec{AC}$:

$$\cos\theta = \frac{b^2 + c^2 - a^2}{2bc}$$

$$\cos\theta = \frac{5^2 + 13^2 - (2\sqrt{65})^2}{2 \times 5 \times 13}$$

$$= -\frac{33}{65}$$

So the angle between $\vec{AB}$ and $\vec{AC}$ is 120.5° (to 1 decimal place).

EXERCISE 10C

TARGETING DISTINCTION

1. The diagram shows a triangle with an extended straight line ABE.

    OA = 2OC

    AB : AE = 1 : 2

    $\vec{OA} = 6\mathbf{a}$ $\vec{OB} = 6\mathbf{b}$

    $\vec{OD} = k\mathbf{b}$, where k is a scalar quantity.

    $\vec{CE} = 3\vec{CD}$

    Given that CDE is a straight line, show that $k = 4$.

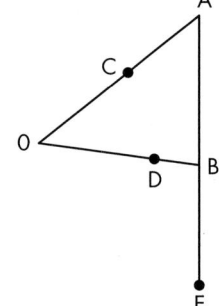

2. The diagram shows a triangle with $\vec{OA} = 3\mathbf{a}$ and $\vec{OB} = 2\mathbf{b}$.

    X is the point on AB such that AX : BX = 4 : 1 and $\vec{OX} = k(3\mathbf{a} + 8\mathbf{b})$.
    Find the value of k.

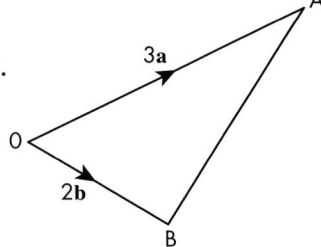

3. The diagram below shows a triangle PQR with a line extended from Q to T.

   $\overrightarrow{PQ} = 4\mathbf{a}$, $\overrightarrow{PR} = 3\mathbf{b}$ and $\overrightarrow{QT} = \frac{4}{3}\mathbf{a} + 8\mathbf{b}$

   S is the point on QR such that QS : SR = 2 : 1.

   Prove that points P, S and T are collinear.

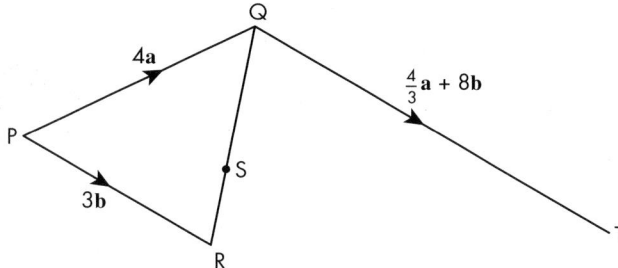

4. Find the angle between the vectors $\overrightarrow{XY} = \begin{pmatrix} 3 \\ -1 \end{pmatrix}$ and $\overrightarrow{XZ} = \begin{pmatrix} -1 \\ 4 \end{pmatrix}$ correct to 1 decimal place.

5. Find the angle between the vectors $\overrightarrow{OA} = \begin{pmatrix} 4 \\ 3 \end{pmatrix}$ and $\overrightarrow{OB} = \begin{pmatrix} 7 \\ -5 \end{pmatrix}$ correct to 1 decimal place.

TARGETING DISTINCTION*

6. The diagram shows a quadrilateral ABCE.

   $\overrightarrow{AE} = \mathbf{a}$, $\overrightarrow{EC} = \mathbf{b}$ and $\overrightarrow{AB} = \mathbf{b} - \mathbf{a}$

   D is the midpoint of CE

   F is the point on BD such that BF : FD = n : 1

   A, F and C are collinear.

   Find the value of n.

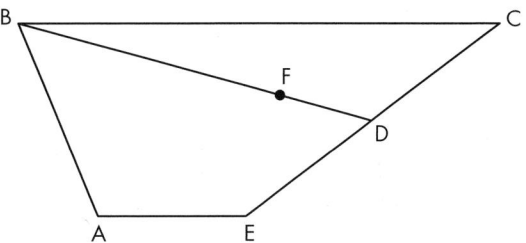

7. The diagram shows a triangle QPS.

   PXR and TXQ are straight lines and R is the midpoint of QS.

   $\overrightarrow{PQ} = \mathbf{a}$ and $\overrightarrow{PS} = \mathbf{b}$

   2PX = 3XR

   Work out the ratio PT : TS.

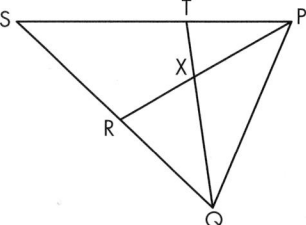

8. VWXYZ is a pentagon.

   Prove by argument that WXYZ is a parallelogram.

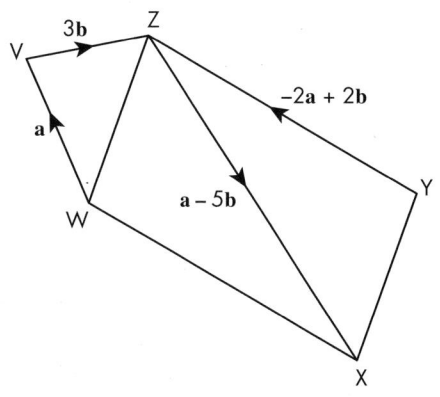

Exam-style questions

1. **a** and **b** are vectors connecting points W, X, Y and Z.

   $\overrightarrow{XY} = \mathbf{a} + \mathbf{b}$ $\overrightarrow{YZ} = 2\mathbf{a} + \mathbf{b}$ $\overrightarrow{ZW} = \mathbf{a} + 2\mathbf{b}$

   a) Show that YW is parallel to XY. [2 marks]

   b) Write down the ratio YW : XY. [1 mark]

c) What do your answers to **a** and **b** tell you about the points X, Y and W? [1 mark]

d) O is the origin. A, B and C are three points such that:

$$\overrightarrow{OA} = \begin{pmatrix} 6 \\ 2 \end{pmatrix} \quad \overrightarrow{OB} = \begin{pmatrix} 1 \\ 1 \end{pmatrix} \quad \overrightarrow{OC} = \begin{pmatrix} 2 \\ -4 \end{pmatrix}$$

Prove that angle ABC is a right angle. [3 marks]

2. The points A and B have position vectors $\begin{pmatrix} -5 \\ 3 \end{pmatrix}$ and $\begin{pmatrix} 4 \\ -2 \end{pmatrix}$, respectively, relative to a fixed origin O.

   Point X is on the line AB and AX : XB = 1 : 2.
   Find the position vector of X. [4 marks]

3. The diagram shows a parallelogram.

   $\overrightarrow{OV} = 2\mathbf{a}$ and $\overrightarrow{OX} = 2\mathbf{b}$
   Y is the point on $\overrightarrow{OW}$ such that $\overrightarrow{OY} = 2\overrightarrow{YW}$.
   Z is the point on $\overrightarrow{XW}$ such that $\overrightarrow{XZ} = \frac{1}{2}\overrightarrow{XW}$.
   Show that V, Y and Z are collinear.

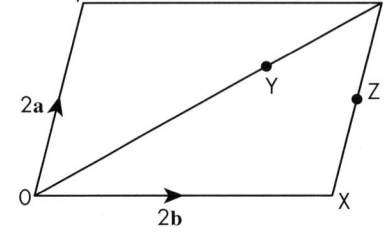

[4 marks]

4. In the diagram above OXZ is a triangle.
   WVZ and OVY are straight lines.
   $\overrightarrow{OZ} = \mathbf{a}$ and $\overrightarrow{OX} = \mathbf{b}$
   XZ = 2XY
   OV = $\frac{3}{5}$OY
   Work out the ratio OW : WX.

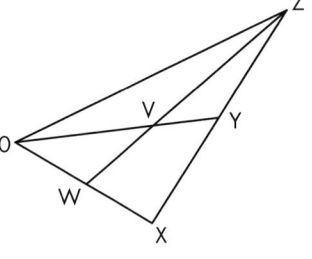

[5 marks]

5. ABC is a triangle. $\overrightarrow{AB} = \mathbf{a}$ and $\overrightarrow{AC} = \mathbf{b}$
   D is the point on AB such that AD : DB = 2 : 1.
   E is the point on BC such that BE : EC = 2 : 5.
   F is the midpoint of AC.
   Find, in terms of **a** and **b**,
   a) $\overrightarrow{FD}$ b) $\overrightarrow{DE}$

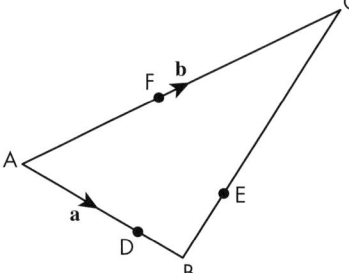

[5 marks]

Answers

CHAPTER 1 NUMBER

1.1 Indices

Exercise 1A

1. a) 5^4 b) 5^3 c) 5^6 d) 5^3 e) 5^5
   f) 6^3 g) 6^3 h) 6^2 i) 6^7 j) 6^1
2. a) 4^6 b) 4^{15} c) 4^6 d) 4^6
   e) 4^8 f) $4^0 = 1$

Exercise 1B

1. a) $\frac{1}{5^3}$ b) $\frac{1}{6}$ c) $\frac{1}{10^5}$ d) $\frac{1}{3^2}$ e) $\frac{1}{8^2}$
   f) $\frac{1}{9}$
2. a) 3^{-2} b) 6^{-1} c) 10^{-3}
3. a) i 2^4 ii 2^{-1} iii 2^{-4} iv -2^3
   b) i 10^3 ii 10^{-1} iii 10^{-2} iv 10^6
   c) i 5^3 ii 5^{-1} iii 5^{-2} iv 5^0
   d) i 3^2 ii 3^{-3} iii 3^0 iv -3^5
4. a) $\frac{49}{9} = 5\frac{4}{9}$ b) $\frac{125}{8} = 15\frac{5}{8}$ c) 4

Exercise 1C

1. a) 5 b) 10 c) 8 d) 9 e) 25
   f) 3 g) 4 h) 10 i) 5 j) 8
   k) 12 l) 20 m) 5 n) 3 o) 10
   p) 3 q) 2 r) 2 s) 6 t) 6
   u) $\frac{1}{4}$ v) $\frac{1}{2}$ w) $\frac{1}{3}$ x) $\frac{1}{5}$ y) $\frac{1}{10}$
2. a) $\frac{5}{6}$ b) $\frac{10}{6} = \frac{5}{3} = 1\frac{2}{3}$ c) $\frac{8}{9}$ d) $\frac{9}{5} = 1\frac{4}{5}$
   e) $\frac{5}{8}$ f) $\frac{3}{5}$ g) $\frac{2}{8} = \frac{1}{4}$ h) $\frac{10}{4} = \frac{5}{2} = 2\frac{1}{2}$
   i) $\frac{4}{5}$ j) $\frac{8}{7} = 1\frac{1}{7}$
3. $64^{-\frac{1}{2}} = \frac{1}{8}$, others are both $\frac{1}{2}$.
4. a) 3 b) $\frac{1}{3}$ c) 0 d) $\frac{1}{2}$ e) $\frac{1}{2}$
   f) $\frac{1}{4}$ g) $\frac{1}{4}$ h) $\frac{1}{3}$ i) $\frac{1}{3}$ j) $\frac{1}{2}$
   k) $\frac{1}{3}$ l) $\frac{1}{7}$

Exercise 1D

1. a) 16 b) 25 c) 216 d) 81
2. a) $\frac{1}{5}$ b) $\frac{1}{6}$ c) $\frac{1}{2}$ d) $\frac{1}{3}$
   e) $\frac{1}{4}$ f) $\frac{1}{2}$ g) $\frac{1}{2}$ h) $\frac{1}{3}$
3. a) $\frac{1}{125}$ b) $\frac{1}{216}$ c) $\frac{1}{8}$ d) $\frac{1}{27}$
   e) $\frac{1}{256}$ f) $\frac{1}{4}$ g) $\frac{1}{4}$ h) $\frac{1}{9}$
4. a) $\frac{1}{100000}$ b) $\frac{1}{12}$ c) $\frac{1}{25}$ d) $\frac{1}{27}$
   e) $\frac{1}{32}$ f) $\frac{1}{32}$ g) $\frac{1}{81}$ h) $\frac{1}{13}$
5. $8^{-\frac{2}{3}} = \frac{1}{4}$, others are both $\frac{1}{8}$
6. a) $\frac{27}{8}$ b) $\frac{9}{25}$ c) $\frac{1024}{243}$ d) $\frac{8}{343}$
   e) $\frac{16}{9}$ f) $\frac{8}{27}$ g) $\frac{625}{256}$ h) $\frac{32}{243}$
7. a) $\frac{25}{9}$ b) $\frac{27}{64}$ c) $\frac{125}{729}$ d) $\frac{243}{32}$
   e) $\frac{16}{25}$ f) $\frac{512}{125}$ g) $\frac{243}{32}$ h) $\frac{32}{243}$

1.2 Surds

Exercise 1E

1. a) $\sqrt{14}$ b) 6 c) 6 d) $\sqrt{30}$
2. a) 2 b) $\sqrt{6}$ c) 1 d) 3
3. a) $4\sqrt{2}$ b) $10\sqrt{2}$ c) $10\sqrt{10}$ d) $5\sqrt{10}$
   e) $7\sqrt{2}$ f) $9\sqrt{3}$
4. a) $2\sqrt{3}$ b) $4\sqrt{2}$ c) $8\sqrt{5}$
5. a) $\sqrt{3}$ b) 1 c) $2\sqrt{2}$ d) $\sqrt{5}$ e) $\sqrt{3}$
   f) $\sqrt{2}$
6. a) a b) 1 c) $\sqrt{a}$
7. a) 36 b) $16\sqrt{30}$ c) 54 d) 32
   e) $20\sqrt{6}$ f) 24 g) 16 h) 18
8. a) 6 b) $3\sqrt{5}$ c) $6\sqrt{6}$ d) $2\sqrt{3}$
9. a) $2\sqrt{3}$ b) 4 c) $6\sqrt{2}$ d) $4\sqrt{2}$
   e) $3\sqrt{2}$ f) $\sqrt{7}$
10. a) abc b) $\frac{a}{c}$ c) $c\sqrt{b}$
11. a) 20 b) 24 c) 10 d) 24
12. a) $\frac{3}{4}$ b) $8\frac{1}{3}$ c) $\frac{5}{16}$ d) 12 e) 2
13. a) False b) False
14. For example, $\sqrt{2} \times \sqrt{8} = 4$
15. a) $5\sqrt{2}$ b) $2\sqrt{3}$ c) $14\sqrt{2}$

1.3 Rationalising denominators

Exercise 1F

1. a) $(2+\sqrt{3})(1+\sqrt{3}) = 2 + 2\sqrt{3} + \sqrt{3} + \sqrt{9}$
   $= 2 + 3\sqrt{3} + 3$
   $= 5 + 3\sqrt{3}$
   b) $(1+\sqrt{2})(2+\sqrt{3}) = 2 + 2\sqrt{2} + \sqrt{3} + \sqrt{6}$

c) $(4-\sqrt{3})(4+\sqrt{3}) = 16 + 4\sqrt{3} - 4\sqrt{3} - \sqrt{9}$
   $= 16 - 3$
   $= 13$

2. a) $2\sqrt{3} - 3$ b) $3\sqrt{2} - 8$ c) $10 + 4\sqrt{5}$
   d) $12\sqrt{7} - 42$ e) $15\sqrt{2} - 24$ f) $9 - \sqrt{3}$

3. a) $2\sqrt{3}$ b) $1 + \sqrt{5}$ c) $\sqrt{7} - 30$
   d) -41 e) $9 + 4\sqrt{5}$ f) $3 - 2\sqrt{2}$

4. a) $(\sqrt{3} - 1)$ cm²   b) $(2\sqrt{5} + 5\sqrt{2})$ cm²
   c) $(2\sqrt{3} + 18)$ cm²

5. a) $\frac{\sqrt{3}}{3}$ b) $\frac{\sqrt{3}}{6}$ c) $\sqrt{3}$
   d) $\frac{3}{2}$ e) $\frac{\sqrt{21}}{3}$ f) $\frac{2\sqrt{3}-3}{3}$
   g) $\frac{5\sqrt{3}+6}{3}$

6. a) i 1   ii -4 iii 2
   b) They become whole numbers. Difference of two squares makes the 'middle terms' (and surds) disappear.

7. Possible answer: $80^2 = 6400$, so $80 = \sqrt{6400}$ and $10\sqrt{70} = \sqrt{7000}$
   Since $6400 < 7000$, there is not enough cable.

8. $9 + 6\sqrt{2} + 2 - (1 - 2\sqrt{8} + 8)$
   $= 11 - 9 + 6\sqrt{2} + 4\sqrt{2}$
   $= 2 + 10\sqrt{2}$

9. $x^2 - y^2 = (1+\sqrt{2})^2 - (1-\sqrt{8})^2$
   $= 1 + 2\sqrt{2} + 2 - (1 - 2\sqrt{8} + 8)$
   $= 3 - 9 + 2\sqrt{2} + 4\sqrt{2}$
   $= -6 + 6\sqrt{2}$
   $(x+y)(x-y) = (2 - \sqrt{2})(3\sqrt{2})$
   $= 6\sqrt{2} - 6$

10. a) $\frac{3\sqrt{2}-1}{17}$ b) $\frac{15+5\sqrt{3}}{6}$ c) $\frac{11+6\sqrt{2}}{7}$
    d) $13 - 5\sqrt{5}$ e) $-5 - 2\sqrt{6}$

11. a) $x = 23, y = 9$ b) $x = 128, y = 648$
    c) $x = \frac{1}{2}, y = -\frac{1}{2}$ d) $x = 2, y = -6$ or $x = -2, y = 6$

Exam-style questions

1. a) 80
   b) $\sqrt{8} - \sqrt{16} + 1 - \sqrt{2} = 2\sqrt{2} - 4 + 1 - \sqrt{2} = -3 + \sqrt{2}$

2. a) 4   b) 4

3. $81\sqrt{3} = 3^4 \times 3^{\frac{1}{2}} = 3^{4\frac{1}{2}} = 3^{\frac{9}{2}} \Rightarrow x = \frac{9}{2}$

4. $(5 + \sqrt{2})(5 - \sqrt{2}) = 25 - 2 = 23$

5. a) $13 - 6\sqrt{5}$ b) $\frac{3}{11} - \frac{2}{11}\sqrt{5}$

6. a) $\frac{1}{1000}$ b) 100   c) $\left(\frac{8}{27}\right)^{\frac{2}{3}} = \left(\frac{2}{3}\right)^2 = \frac{4}{9}$

7. $\frac{1-\sqrt{3}+2+\sqrt{3}}{(2+\sqrt{3})^2} = \frac{3}{4+4\sqrt{3}+3}$
   $= \frac{3}{7+4\sqrt{3}} \times \frac{7-4\sqrt{3}}{7-4\sqrt{3}}$
   $= \frac{3(7-4\sqrt{3})}{49-48}$
   $= 3(7 - 4\sqrt{3})$

8. $\frac{4+\sqrt{2}}{3-\sqrt{2}} \times \frac{3+\sqrt{2}}{3+\sqrt{2}} = \frac{12+4\sqrt{2}+3\sqrt{2}+2}{9-2}$
   $= \frac{14+7\sqrt{2}}{7}$
   $= 2 + \sqrt{2}$
   $P = 2(2+\sqrt{2}) + 2(3-\sqrt{2})$
   $= 4 + 2\sqrt{2} + 6 - 2\sqrt{2}$
   $= 10$

9. $\left(\frac{9}{\sqrt{27}}\right)^2 = \frac{81}{27} = 3$
   $\left(\frac{243}{3125}\right)^{-\frac{2}{5}} = \frac{25}{9} = 2\frac{7}{9}$, so $\left(\frac{9}{\sqrt{27}}\right)^2$ is bigger.

CHAPTER 2 ALGEBRAIC MANIPULATION
2.1 Algebraic indices
Exercise 2A

1. a) $12c^3$ b) $4k^6$ c) $3y^{-1}$ d) w^{-2} e) $8d^5$
   f) $2d^{-7}$ g) $6a^{-1}$ h) $4p^{-7}$ i) $\frac{1}{3}r^{-7}$

2. a) $t^{\frac{2}{3}}$ b) $m^{\frac{3}{4}}$ c) $k^{\frac{2}{5}}$ d) $h^{-\frac{3}{2}}$

3. a) x^4 b) x^{-1} c) $4y^2$ d) $10x^2$
   e) $20x^{-1}$ f) $\frac{1}{3}y$

4. a) x b) d^{-1} c) $t^{\frac{3}{2}}$ d) x^2
   e) $y^{\frac{1}{2}}$ f) a^4

5. a) $x^{\frac{1}{2}}$ b) y^{-1} c) $a^{\frac{5}{3}}$ d) t^{-2}
   e) d^2 f) 1

6. a) $2a$ b) $4a^{-5}$ c) $2a^{\frac{3}{2}}$

7. a) $5 - 6x^{-2}$ b) $3x^6 - 2x$

8. a) x^3 b) x^{-1} c) x d) x e) $x^{\frac{3}{2}}$ f) x

9. a) $\frac{16}{3}x$ b) $\frac{1}{3}x^{\frac{1}{3}}$ c) $8x^{\frac{1}{2}}$

2.2 Expanding brackets
Exercise 2B

1. $(3x-2)(2x+1) = 6x^2 - x - 2$,
   $(2x-1)(2x-1) = 4x^2 - 4x + 1$,
   $(6x-3)(x+1) = 6x^2 + 3x - 3$,
   $(4x+1)(x-1) = 4x^2 - 3x - 1$,
   $(3x+2)(2x+1) = 6x^2 + 7x + 2$

2. a) $4x^2 - 1$ b) $25y^2 - 9$ c) $16h^2 - 1$
   d) $4 - 9x^2$ e) $36 - 25y^2$ f) $a^2 - b^2$
   g) $4m^2 - 9p^2$ h) $a^2b^2 - c^2d^2$ i) $a^4 - b^4$
3. First shaded area is $(2k)^2 - 1^2 = 4k^2 - 1$
   Second shaded area is $(2k+1)(2k-1) = 4k^2 - 1$
4. a) $x^3 + 4x^2 + 2x - 1$ b) $8x^3 + 22x^2 + 9x + 1$
   c) $2x^4 - 9x^3 - 14x^2 - 9$ d) $x^6 + x^5 - 6x^4 + 4x^3 - x$
5. a) $x^3 + 3x^2 + 3x + 1$ b) $8x^3 - 12x^2 + 6x - 1$
   c) $27x^3 + 54x^2 + 36x + 8$ d) $64x^3 - 144x^2 + 108x - 27$
6. a) $2x - 2$ b) $x - 1$
   c) $6x - 13 + \frac{6}{x}$ d) $16x - x^{-\frac{2}{3}}$

Exercise 2C

1. $x^4 - 12x^3 + 54x^2 - 108x + 81$
2. 5000
3. -6144
4. 2916
5. 2
6. 3
7. $1 + 2x + \frac{3}{2}x^2 + \frac{1}{2}x^3 + \frac{1}{16}x^4$
8. $p = 5$

2.3 Factorising

Exercise 2D

1. $(y+6)(y-1)$
2. $(m+2)(m-6)$
3. $(n+3)(n-6)$
4. $(m+4)(m-11)$
5. $(t+9)(t-10)$
6. $(h+8)(h-9)$
7. $(t+7)(t-9)$
8. $(y+10)^2$
9. $(m-9)^2$
10. $(x-12)^2$
11. $(d+3)(d-4)$
12. $(q+7)(q-8)$
13. $(x+2y)(x-2y)$
14. $(3x+1)(3x-1)$
15. $(4x+3)(4x-3)$
16. $(5x+8y^2)(5x-8y^2)$
17. $(\frac{1}{2}x + \frac{1}{3}y)(\frac{1}{2}x - \frac{1}{3}y)$
18. $(\frac{2}{9}x + 10)(\frac{2}{9}x - 10)$

Exercise 2E

1. $(2x+1)(x+2)$
2. $(7x+1)(x+1)$
3. $(4x+7)(x-1)$
4. $(3t+2)(8t+1)$
5. $(3t+1)(5t-1)$
6. $(4x-1)^2$
7. $3(y+7)(2y-3)$
8. $4(y+6)(y-4)$
9. $(2x+3)(4x-1)$
10. $(2t+1)(3t+5)$
11. $(3x+2)(x-6)$
12. $-(7x-2)(x-5)$
13. $4x+1$ and $3x+2$
14. a) All the terms in the quadratic have a common factor of 6, but this has not been taken out, so these are not complete factorisations. 36 has several different pairs of factors which all work.
   b) $6(x+2)(x+3)$. This has the highest common factor taken out.

Exercise 2F

1. $(5x-y)(3x+y)$
2. $(x^2+5y^2)(x^2-5y^2)$
3. $2x(2x+5)(2x-5)$
4. $(4x^2+5y^2)(4x^2-5y^2)$
5. $16(x+1)$
6. $8x$
7. $5(x+1)(x-1)$
8. $-(4x+1)$ or $-4x-1$
9. $(8x+1)(2x+1)$
10. $(2x^2-3y)(7x^2+2y)$
11. $2x^3(3x-y)(x-2y)$
12. $(\frac{1}{2}p^2+q)(\frac{1}{2}p^2+2q)$

Exercise 2G

1. a) $(-1)^3 + 6(-1)^2 - 9(-1) - 14 = -1 + 6 + 9 - 14 = 0$
   b) $(3)^3 + 3(3)^2 - 13(3) - 15 = 27 + 27 - 39 - 15 = 0$
   c) $(4)^3 - 7(4)^2 + 2(4) + 40 = 64 - 112 + 8 + 40 = 0$
   d) $(-6)^3 + 13(-6)^2 + 54(-6) + 72 = -216 + 468 - 324 + 72 = 0$
   e) $(-7)^3 - 37(-7) + 84 = -343 + 259 + 84 = 0$
   f) $2(1.5)^3 - 5(1.5)^2 + (1.5) + 3 = 6.75 - 11.25 + 1.5 + 3 = 0$
2. a) $(x+1)(x-3)(x+5)$ b) $(x+4)(x-1)(x+2)$
   c) $(x+1)(x+2)(x+3)$ d) $(x-2)(x+3)(x+5)$
   e) $(2x+1)(x-4)(x-3)$ f) $(3x-1)(2x+3)(x+2)$
3. $(x+2)$
4. a) $(-5)^3 + 3(-5)^2 - 13(-5) + c = 0 \Rightarrow -125 + 75 + 65 + c = 0 \Rightarrow c = -15$
   b) $x+1, x-3$

5. $a = 3, b = 1, c = 1$
6. $x - 4, x + 4, x + 3$
7. $x + 2$ and $x - 4$
8. $f(-3) = (-3)^4 - 13(-3)^2 + 36 = 0$
9. $f\left(\frac{5}{2}\right) = 2\left(\frac{5}{2}\right)^4 + \left(\frac{5}{2}\right)^3 - 17\left(\frac{5}{2}\right)^2 - \frac{5}{2} + 15 = 0$
10. a) $f\left(\frac{1}{3}\right) = 3\left(\frac{1}{3}\right)^3 - 22\left(\frac{1}{3}\right)^2 + 43\left(\frac{1}{3}\right) - 12 = 0$
    b) $(3x - 1)(x - 3)(x - 4)$

2.4 Completing the square

Exercise 2H

1. a) $(x + 2)^2 - 4$ b) $(x + 7)^2 - 49$
   c) $(x - 3)^2 - 9$ d) $(x + 3)^2 - 9$
   e) $(x - 1.5)^2 - 2.25$ f) $(x - 4.5)^2 - 20.25$
   g) $(x + 6.5)^2 - 42.25$ h) $(x + 5)^2 - 25$
   i) $(x + 4)^2 - 16$ j) $(x - 1)^2 - 1$
   k) $(x + 1)^2 - 1$ l) $\left(x - \frac{5}{2}\right)^2 - \frac{25}{4}$

2. a) $(x + 2)^2 - 5$ b) $(x + 7)^2 - 54$
   c) $(x - 3)^2 - 6$ d) $(x + 3)^2 - 2$
   e) $(x - 1.5)^2 - 3.25$ f) $\left(x - \frac{9}{2}\right)^2 - \frac{41}{4}$
   g) $\left(x - \frac{13}{2}\right)^2 - \frac{29}{4}$ h) $(x + 5)^2 - 22$
   i) $(x + 4)^2 - 22$ j) $(x + 1)^2 - 2$
   k) $(x - 1)^2 - 8$ l) $(x + 1)^2 - 10$

3. a) $2(x + 1)^2 + 5$ b) $3(x + 2)^2 - 9$
   c) $6(x + 1)^2 - 2$ d) $5(x - 3)^2 - 33$
   e) $8(x - 2)^2 - 22$ f) $9(x + 0.5)^2 + 6.75$
   g) $12(x - 1.5)^2 - 13$ h) $5(x + 1)^2 + 1$
   i) $7(x + 1)^2 - 2$ j) $7(x + 0.5)^2 + 0.25$
   k) $10(x - 1)^2 - 5$ l) $11(x + 1)^2 - 5$

4. $a = 4, b = 4, c = 9$
5. $a = 24, b = 4, c = -41$
6. $a = 9, b = 2$
7. $p = -14, q = -3$

2.5 Algebraic fractions

Exercise 2I

1. a) $\frac{x^2y + 8}{4x}$ b) $\frac{7x + 3}{4}$ c) $\frac{13x + 5}{15}$
   d) $\frac{5x - 10}{4}$ e) $\frac{xy^2 - 8}{4y}$ f) $\frac{x + 1}{4}$
   g) $\frac{-7x - 5}{15}$ h) $\frac{2 - 3x}{4}$ i) $\frac{2xy}{3}$
   j) $\frac{x^2 - 2x}{10}$ k) $\frac{1}{6}$ l) $\frac{1}{2x}$
   m) x n) 3 o) $\frac{13x + 9}{10}$
   p) $\frac{x + 3}{2}$

2. a) $\frac{7x + 9}{(x + 1)(x + 2)}$ b) $\frac{11x - 10}{(x - 2)(x + 1)}$
   c) $\frac{x + 1}{(2x - 1)(3x - 1)}$

3. a) $\frac{9x + 13}{(x + 1)(x + 2)}$ b) $\frac{14x + 19}{(4x - 1)(x + 1)}$
   c) $\frac{2x^2 + x - 13}{2(x + 1)}$

4. $\frac{2x^2 + x - 3}{4x^2 - 9}$

5. a) $\frac{x - 1}{2x + 1}$ b) $\frac{2x + 1}{x + 3}$ c) $\frac{2x - 1}{3x - 2}$
   d) $\frac{x + 1}{x - 1}$ e) $\frac{2x + 5}{4x - 1}$

6. a) $\frac{x(x - 1)}{x - 2}$ b) $\frac{x(x - 5)}{x - 4}$ c) $\frac{x(x - 4)}{x - 5}$
   d) $\frac{x(x - 6)}{x - 7}$ e) $\frac{x(x - 5)}{x + 3}$ f) $\frac{x(x + 3)}{x + 8}$

7. a) $x + 2$ b) $x + 3$ c) $4x + 5$
   d) $2x - 7$ e) $3(x + 1)$ f) $x + 2$

8. a) $\frac{4x(3x + 1)}{2x - 3}$ b) $\frac{x(9x - 8)}{3x - 7}$ c) $\frac{2x(x + 4)}{4x - 3}$
   d) $\frac{6(x - 1)}{5x + 6}$ e) $\frac{6x(x + 1)}{8x - 3}$ f) $\frac{x(4x - 5)}{3x + 4}$

9. $\frac{2x(7x - 3)(x - 5)}{(7x - 3)(x + 4)} = \frac{2x(x - 5)}{(x + 4)}$

Exam-style questions

1. $26y$
2. x
3. $x^{\frac{1}{6}}$
4. $3^{3a + 2}$
5. $3x^{\frac{1}{2}} - x + 1$
6. $7(d - 2)$
7. a) $64 - 48x + 12x^2 - x^3$ b) $x + 1$
8. Coefficient $10(2)^3(-a)^2 = 720, 80a^2 = 720, a^2 = 9$
   so, as a is positive, $a = 3$
9. $(4x + 1)^2 - 7x(x + 1) + 5x = 16x^2 + 8x + 1 - 7x^2 - 7x + 5x = 9x^2 + 6x + 1 = (3x + 1)^2 \geq 0$
10. a) $(3x + 2)(3x - 2)$ b) $(3x - 2)(2x + 1)$
11. $16(x + 1)(x - 1)$
12. $-6x^2(2x - 3)(x + 4)$
13. $a = 6, b = 0.5, c = -2.5$
14. $\frac{3}{(x + 1)(x + 2)}$
15. $\frac{(2x - 1)(x - 3)}{2x + 1}$
16. $\frac{x}{(x + 2)^2}$
17. a) $f(-4) = 0$ b) $(x + 4)(x - 3)(x - 2)$
18. $(x - 4)(x - 5)(x + 5)$
19. a) $f\left(-\frac{1}{2}\right) = 0$ b) $(2x + 1)(x - 3)(x + 2)$

CHAPTER 3 GRAPHS

3.1 Linear graphs

Exercise 3A

1. **a)** $m = \frac{5--3}{4-0} = \frac{8}{4} = 2$
   $y - 5 = 2(x - 4) \Rightarrow y - 5 = 2x - 8 \Rightarrow y = 2x - 3$

   b) $m = \frac{5-2}{2--4} = \frac{3}{6} = \frac{1}{2}$
   $y - 5 = \frac{1}{2}(x - 2) \Rightarrow y - 5 = \frac{1}{2}x - 1 \Rightarrow y = \frac{1}{2}x + 4$

2. **a)** $2x - y - 3 = 0$ **b)** $x - 2y + 8 = 0$

3. $m = \frac{4-q-(-7q)}{p-2-3p} = \frac{4+6q}{-2-2p} = -\frac{2+3q}{1+p}$

4. **a)** $y = 4x + 1$ **b)** $y = \frac{1}{2}x - 2$

5. **a)** $y = -\frac{1}{3}x - 1$ **b)** $y = -3x + 5$

6. **a)** $y = -\frac{1}{2}x - 1$ **b)** $y = \frac{1}{3}x + 1$
   c) $y = -2x + 3$ **d)** $y = \frac{3}{2}x - 5$

7. $m = -\frac{1}{4}$
   $y - 3 = -\frac{1}{4}(x + 4) \Rightarrow y - 3 = -\frac{1}{4}x - 1 \Rightarrow y = -\frac{1}{4}x + 2$

8. **a)** $m = \frac{5-3}{1-3} = \frac{2}{-2} = -1$
   $y - 9 = -(x - 5) \Rightarrow y - 9 = -x + 5 \Rightarrow y = -x + 14$

   b) $m = 1$
   Midpoint $= \left(\frac{1+3}{2}, \frac{5+3}{2}\right) = \left(\frac{4}{2}, \frac{8}{2}\right) = (2, 4)$
   $y - 4 = x - 2 \Rightarrow y = x + 2$

9. $m_1 = \frac{6-2}{3-1} = \frac{4}{2} = 2$
   $m_2 = -\frac{1}{2}$
   Midpoint $= \left(\frac{1+3}{2}, \frac{2+6}{2}\right) = \left(\frac{4}{2}, \frac{8}{2}\right) = (2, 4)$
   $y - 4 = -\frac{1}{2}(x - 2) \Rightarrow y - 4 = -\frac{1}{2}x + 1 \Rightarrow y = -\frac{1}{2}x + 5$

10. Line A: $\frac{3-0}{4--3} = \frac{3}{7}$, line B: $\frac{-12--6}{-13-1} = \frac{-6}{-14} = \frac{3}{7}$
    Lines A and B are parallel as their gradients are equal.

11. $y = -3x + 10 \Rightarrow m_1 = -3$
    $m_2 = \frac{1}{3}$
    $y - 5 = \frac{1}{3}(x - 3) \Rightarrow y - 5 = \frac{1}{3}x - 1 \Rightarrow y = \frac{1}{3}x + 4$

12. $m = -\frac{3}{2} \Rightarrow m_2 = \frac{2}{3}$
    $y - 3 = \frac{2}{3}(x + 2) \Rightarrow y - 3 = \frac{2}{3}x + \frac{4}{3} \Rightarrow y = \frac{2}{3}x + \frac{13}{3}$ or
    $2x - 3y + 13 = 0$

13. $m_1 = \frac{7--5}{5-2} = \frac{12}{3} = 4$
    $m_2 = -\frac{1}{4}$
    $m_2 = \frac{p-3}{-7-1} = \frac{p-3}{-8} = -\frac{1}{4} = \frac{2}{-8} \Rightarrow p - 3 = 2 \Rightarrow p = 5$

14. $m_{AB} \frac{2-4}{7-3} = \frac{-2}{4} = -\frac{1}{2}$, $m_{BC} \frac{-3-2}{10-7} = \frac{-5}{3} = -\frac{5}{3}$
    $m_{CD} \frac{1--3}{2-10} = \frac{4}{-8} = -\frac{1}{2}$, $m_{DA} \frac{1-4}{2-3} = \frac{-3}{-1} = 3$
    ABCD is a trapezium as it has one pair of parallel sides.

15. $m_{AB} \frac{-2-6}{6-4} = \frac{-8}{2} = -4$, $m_{BC} \frac{-5--2}{-6-6} = \frac{-3}{-12} = \frac{1}{4}$
    $m_{CD} \frac{3--5}{-8--6} = \frac{8}{-2} = -4$, $m_{DA} \frac{3-6}{-8-4} = \frac{-3}{-12} = \frac{1}{4}$
    ABCD has two pairs of parallel sides: AB is parallel to CD and BC is parallel to AD. AB and CD are perpendicular to BC and AD.
    Therefore, ABCD is a rectangle.

16. $m = \frac{-7-5}{3-(-2)} = \frac{-12}{5} = -\frac{12}{5}$ $m_2 = \frac{5}{12}$
    $y + 7 = \frac{5}{12}(x - 3) \Rightarrow y + 7 = \frac{5}{12}x - \frac{5}{4}$
    $\Rightarrow y = \frac{5}{12}x - \frac{33}{4}$
    $\Rightarrow 5x - 12y = 99$

17. Gradient of radius: $m = \frac{\frac{11}{5}}{\frac{2}{5}} = \frac{11}{2}$
    Tangent: $y = -2x + 3$, $m = -2$
    $m_1 \times m_2 = \frac{11}{2} \times -2 = -11$, so no, the line is not a tangent as the product of the gradients is not -1

18. $m_{AB} = \frac{7-p}{-6--2} = \frac{7-p}{-4}$, $m_{BC} = \frac{3-7}{q--6} = \frac{-4}{q+6}$
    $\frac{7-p}{-4} \times \frac{-4}{q+6} = -1 \Rightarrow 7 - p = -q - 6 \Rightarrow p - q = 13$

19. **a)** $m = \frac{7-1}{12--2} = \frac{6}{14} = \frac{3}{7}$
    $y - 1 = \frac{3}{7}(x + 2) \Rightarrow y - 1 = \frac{3}{7}x + \frac{6}{7} \Rightarrow y = \frac{3}{7}x + \frac{13}{7}$
    $7y = 3x + 13 \Rightarrow 3x - 7y + 13 = 0$

    b) $y - n = -\frac{7}{3}(x - 9) \Rightarrow y - n = -\frac{7}{3}x + 21$
    $\Rightarrow n = \frac{7}{3}x + y - 21$

    c) $n = \frac{7}{3}(-2) + (1) - 21 = -\frac{74}{3}$
    $-\frac{74}{3} = \frac{7}{3}x + y - 21 \Rightarrow -74 = 7x + 3y - 63$
    $\Rightarrow 7x + 3y + 11 = 0$

3.2 Quadratic graphs

Exercise 3B

1. **a)** x-intercepts: $x = -1$, $x = 3$
   y-intercept: $(0, 3)$
   Vertex: $(1, 4)$

   b) x-intercept: $x = -2$
   y-intercept: $(0, 4)$
   Vertex: $(-2, 0)$

   c) No x-intercepts
   y-intercept: $(0, 1.5)$
   Vertex: $(-0.75, 0.38)$

2. a)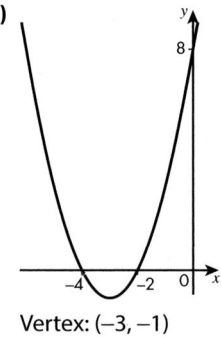
Vertex: (−3, −1)

b)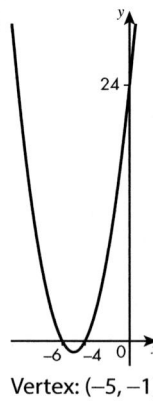
Vertex: (−5, −1)

c)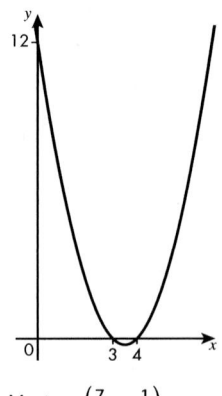
Vertex: $\left(\frac{7}{2}, -\frac{1}{4}\right)$

d)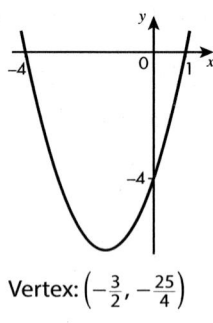
Vertex: $\left(-\frac{3}{2}, -\frac{25}{4}\right)$

3. a)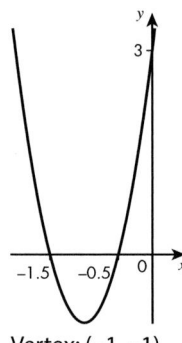
Vertex: (−1, −1)

b)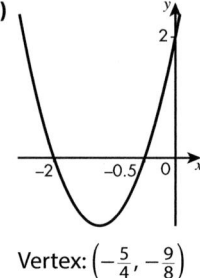
Vertex: $\left(-\frac{5}{4}, -\frac{9}{8}\right)$

c)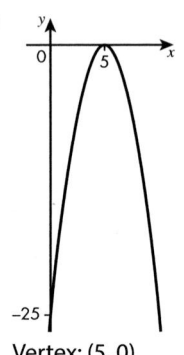
Vertex: (5, 0)

d)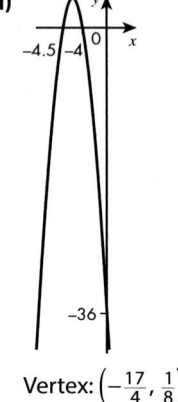
Vertex: $\left(-\frac{17}{4}, \frac{1}{8}\right)$

4. a) x-intercepts at $x = -2, x = -5$
y-intercept: (0, 10)
$y = (x + 2)(x + 5)$
$y = x^2 + 7x + 10$

b) x-intercepts at $x = -2, x = 2$
y-intercept: (0, 4)
$y = (x - 2)(x + 2)$
$y = x^2 - 4$
Negative parabola; therefore
$y = -x^2 + 4$

c) y-intercept: (0, 5)
Vertex: (3, −4)
$y = (x - 3)^2 - 4$
$y = x^2 - 6x + 9 - 4$
$y = x^2 - 6x + 5$

Exercise 3C

1. a) $(-6)^2 - 4(1)(9) = 0$
One real/two equal roots.

b) $(-10)^2 - 4(1)(-24) = 196$
Two real roots.

c) $(-7)^2 - 4(1)(40) = -111$
No real roots.

d) $(6)^2 - 4(2)(-8) = 100$
Two real roots.

e) $(0)^2 - 4(1)(6) = -24$
No real roots.

f) $(24)^2 - 4(3)(48) = 0$
One real/two equal roots.

g) $(9)^2 - 4(-1)(-26) = -23$
No real roots.

h) $(-7)^2 - 4(-2)(12) = 145$
Two real roots.

2. a) $(23)^2 - 4p(6) > 0$
$529 - 24p > 0$
$24p < 529$
$p < \frac{529}{24}$

b) $(7)^2 - 4(2)p > 0$
$49 - 8p > 0$
$8p < 49$
$p < \frac{49}{8}$

c) $(-4)^2 - 4(1) \times 3p > 0$
$16 - 12p > 0$
$12p < 16$
$p < \frac{4}{3}$

3. a) $q^2 - 4(1)(4) = 0$
$q^2 - 16 = 0$
$q^2 = 16$
$q^2 = \pm 4$
$q > 0, \therefore q = 4$

b) $(12)^2 - 4 \times q \times 4q = 0$
$144 - 16q^2 = 0$
$16q^2 = 144$
$q^2 = 9$
$q = \pm 3$
$q > 0, \therefore q = 3$

c) $(-q)^2 - 4(3)\left(\frac{1}{2}q\right) = 0$
$q^2 - 6q = 0$
$q(q - 6) = 0$
$q = 0, q = 6$
$q > 0, \therefore q = 6$

4. a) $(3)^2 - 4(n-1)(-n^2) < 0$
$9 - 4(n^2 - n^3) < 0$
$9 - 4n^2 + 4n^3 < 0$
$4n^3 - 4n^2 < -9$
$4n^2(n-1) < -9$
$n^2(n-1) < -\frac{9}{4}$

3.3 Cubic and quartic graphs

Exercise 3D

1. a) $y = (x-3)(x-1)(x+5)$ **b)** $y = x(x-1)(x+3)$ **c)** $y = 3(x-2)(x+1)(x+3)$

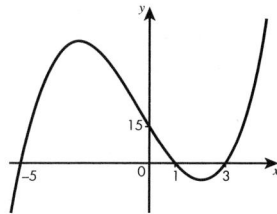 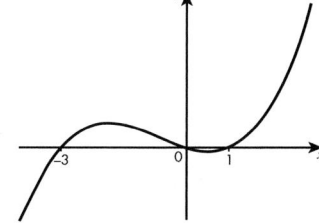 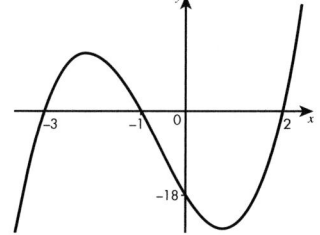

2. a) $y = (x-3)^2(x-1)$ **b)** $y = x^2(x-1)$ **c)** $y = 2(x-2)(x+1)^2$

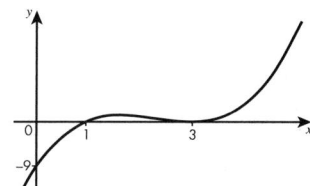  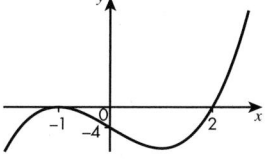

d) $y = -(x-2)^3$

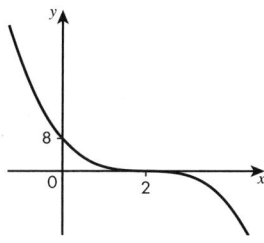

3. a) $y = x^3 - 6x^2 + 3x + 10$
$y = (x+1)(x-2)(x-5)$

b) $y = x^3 - 7x - 6$
$y = (x+2)(x+1)(x-3)$

c) $y = -x^3 - 3x^2 + 10x$
$y = x(x+5)(x-2)$

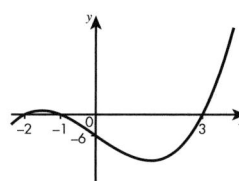 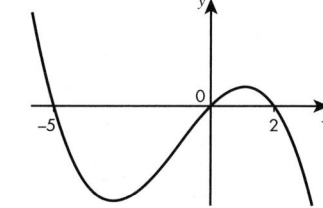

4. $y = (x+3)(x+1)(x-3) = x^3 + x^2 - 9x - 9$
5. $y = (x+2)(x-4)^2 = x^3 - 6x^2 + 32$

Exercise 3E

1. **a)** $y = (x-4)(x-2)(x+1)(x+5)$

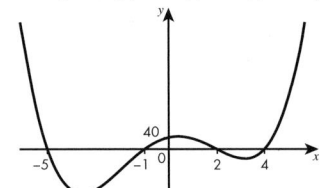

b) $y = x(x-2)(x+1)(x+3)$

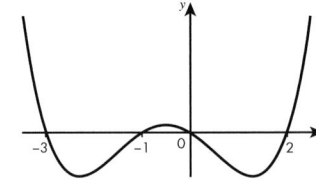

c) $y = 4(x-3)(x-2)(x+1)(x+3)$

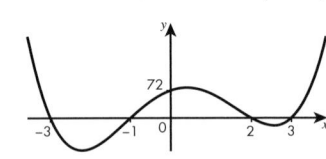

2. **a)** $y = (x-2)^2(x-1)(x+2)$

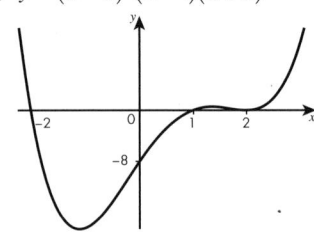

b) $y = x^2(x-3)(x-2)$

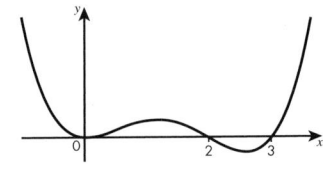

c) $y = (x-1)(x+2)^3$

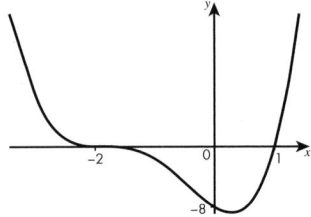

d) $y = (x-2)^4$

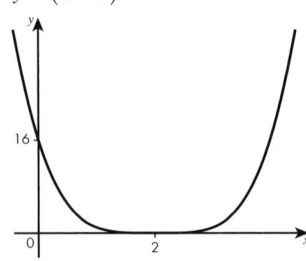

3.4 Trigonometric graphs

Exercise 3F

1. **a)** 36.9°, 143.1° **b)** 53.1°, 126.9°
   c) 48.6°, 131.4° **d)** 224.4°, 315.6°
   e) 194.5°, 345.5

2. sin 234°, as the others all have the same numerical value

3. **a)** 438° or 78° + 360n° **b)** −282° or 78° −360n°
   c) Line symmetry about ±90n° where n is an odd integer. Rotational symmetry about ±180n°, where n is an integer

4. **a)** 53.1°, 306.9° **b)** 54.5°, 305.5°
   c) 62.7°, 297.3° **d)** 143.1°, 216.9°
   e) 104.5°, 255.5°

5. cos 58°, as the others are negative

6. **a)** 492° or 132° + 360n° **b)** −228° or 132° − 360n°
   c) Line symmetry about ±180n° where n is an integer. Rotational symmetry about ±90n°, where n is an odd integer

7. **a)** 14.5°, 194.5° **b)** 38.1°, 218.1°
   c) 50.0°, 230.0° **d)** 160.3°, 340.3°
   e) 147.6°, 327.6°

8. tan 235°, as the others have a numerical value of 1

9. **a)** 425° or 65° + 180n°, where n is an integer.
   b) −115° or 65° − 180n°
   c) No line symmetry. Rotational symmetry about ±180n°, where n is an integer

10. **a)** $x = 21.2°, 158.8°$ **b)** $x = 209.1°, 330.9°$
    c) $x = 50.1°, 309.9°$

11. 30°, 150°

12. **a)** 115° **b)** 327° **c)** 324°

13. **a)** 210°, 330° **b)** 135°, 225°

14. −0.755

15. **a)** $\sin^{-1}(0.5) = 30°; 180° − 30° = 150°$
    $x + 20° = 30°, 150° \Rightarrow x = 10°, 130°$
    b) $\cos^{-1}(0.45) = 63.2°; 360° − 63.2° = 296.7°$
    $5x = 63.2°, 296.7° \Rightarrow x = 12.7°, 59.3°$

16. a) Say 32°, sin 32° = 0.53, cos 58° = 0.53
   b) Say 70°, sin 70° = 0.94, cos 20° = 0.94
   c) sin x = cos $(90 - x)°$
17. a) $90° - 26° = 64°$
   b) $180° + 26° = 206°, 360° - 26° = 334°$
   c) $180° + 64° = 244°, 180° - 64° = 116°$
18. a–e) All true   **f)** False

Exam-style questions

1. a) $m = \frac{7-(-5)}{(-2)-3} = \frac{12}{-5} = -\frac{12}{5}$

$y - 7 = -\frac{12}{5}(x+2) \Rightarrow y - 7 = -\frac{12}{5}x - \frac{24}{5}$

$\Rightarrow y = -\frac{12}{5}x + \frac{11}{5}$

$\Rightarrow 12x + 5y - 11 = 0$

b) $m = \frac{5}{12}$
Midpoint $= \left(\frac{1}{2}, 1\right)$

$y - 1 = \frac{5}{12}\left(x - \frac{1}{2}\right) \Rightarrow y - 1 = \frac{5}{12}x - \frac{5}{24}$

$\Rightarrow y = \frac{5}{12}x + \frac{19}{24}$

2. a) $x = 180° - 27 = 153°$
   b) $y = 180° + 27° = 207°, y = 360° - 27° = 333°$

3. $a\left(x + \frac{2}{3}\right)^2 - \frac{2}{3} \Rightarrow a\left(x^2 + \frac{4}{3}x + \frac{4}{9}\right) - \frac{2}{3}$

$\Rightarrow ax^2 + \frac{4}{3}ax + \frac{4}{9}a - \frac{2}{3}$

$\frac{4}{9}a - \frac{2}{3} = -2 \Rightarrow 4a - 6 = -18$

$\Rightarrow 4a = -12$

$\Rightarrow a = -3$

$y = (-3)x^2 + \frac{4}{3}(-3)x + \frac{4}{9}(-3) - \frac{2}{3}$

$\Rightarrow y = -3x^2 - 4x - 2$

4. $t\left(3 - \frac{1}{2}t\right) = 0$
$t = 0$ (when the ball was thrown)
$3 - \frac{1}{2}t = 0 \Rightarrow 3 = \frac{1}{2}t \Rightarrow t = 6$, 6 seconds.

5. $4^2 \times -2 \times r = -8 \Rightarrow -32r = -8 \Rightarrow r = \frac{1}{4}$
$y = (x+4)^2(x-2)\left(x + \frac{1}{4}\right)$

6. a)

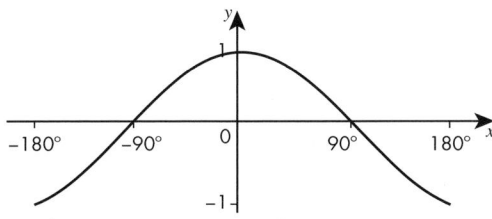

b) $\cos^{-1} 0.85 = 31.8°, x = \pm 31.8°$

7. a) -0.17 **b)** -0.94 **c)** 0.17

8. $y = -\frac{2}{3}x + 2 \Rightarrow y - 4 = -\frac{2}{3}(x-4)$

$\Rightarrow y - 4 = -\frac{2}{3}x + \frac{8}{3}$

$\Rightarrow y = -\frac{2}{3}x + \frac{20}{3}$

9. $\left(2\sqrt{2}q\right)^2 - 4(q+1)(2q-1) = 0$

$\Rightarrow 8q^2 - \left(4\left(2q^2 + q - 1\right)\right) = 0$

$\Rightarrow 8q^2 - \left(8q^2 + 4q - 4\right) = 0$

$\Rightarrow 8q^2 - 8q^2 - 4q + 4 = 0$

$-4q + 4 = 0 \Rightarrow 4q = 4 \Rightarrow q = 1$

10. a)

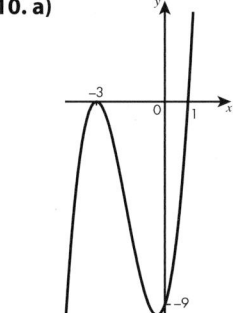

b) $af(x) = a(x+3)(x-1) = a(x^2 + 2x - 3)$
Meets y-axis at $(0, -9)$. $-3a = -9 \Rightarrow a = 3$

11. a) Radius: $m = \frac{6-2}{-1-(-4)} = \frac{4}{3}$

Tangent: $m = -\frac{3}{4} \Rightarrow y - 6 = -\frac{3}{4}(x+1)$

$\Rightarrow y - 6 = -\frac{3}{4}x - \frac{3}{4}$

$\Rightarrow y = -\frac{3}{4}x + \frac{21}{4}$

$\Rightarrow 3x + 4y = 21$

Point A: $\left(0, \frac{21}{4}\right)$,
Point B: $3x + 4(0) = 21 \Rightarrow 3x = 21$

$\Rightarrow x = 7$

$\Rightarrow (7, 0)$

b) Area $= \frac{1}{2} \times \frac{21}{4} \times 7 = \frac{147}{8}$ units² (or 18.375 units²)

CHAPTER 4 MORE GRAPHS

4.1 Translating and reflecting graphs

Exercise 4A

1. a) $y = f(x-2)$

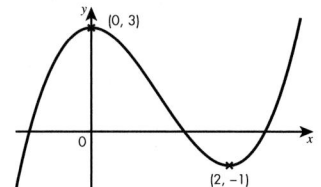

b) $y = f(x) + 1$

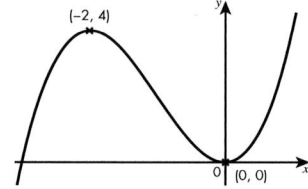

c) $y = f(x) - 3$

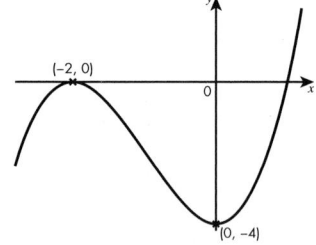

d) $y = f(x+1)$

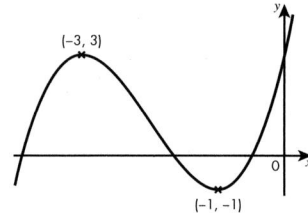

e) $y = f(-x)$

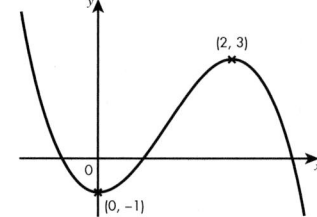

f) $y = -f(x)$

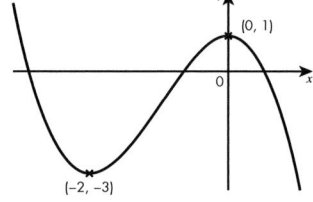

2. a) $A(-4, 2), B(-3, 3), C(1, 7)$ **b)** $A(-1, 0), B(0, 1), C(4, 5)$ **c)** $A(-1, 9), B(0, 10), C(4, 14)$

d) $A(3, 2), B(4, 3), C(8, 7)$ **e)** $A(-1, -2), B(0, -3), C(4, -7)$ **f)** $A(1, 2), B(0, 3), C(-4, 7)$

3. a)

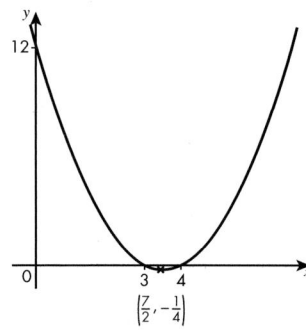

b) i) $y = f(x+3)$

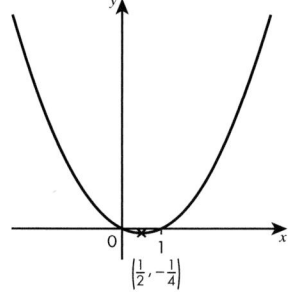

ii) $y = f(x) + 3$

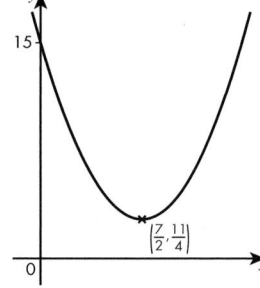

iii) $y = -f(x)$

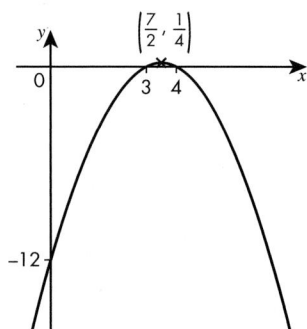

iv) $y = f(-x)$

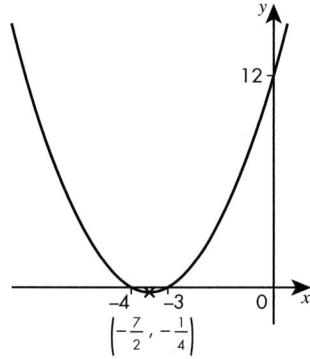

4. $y = f(x - 5.5) - 5.5$

5. a) $y = f(x - 90°)$
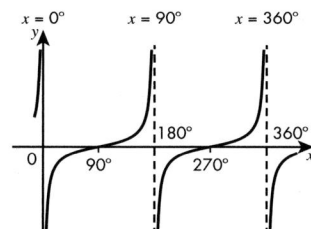

b) $y = f(x) + 1$
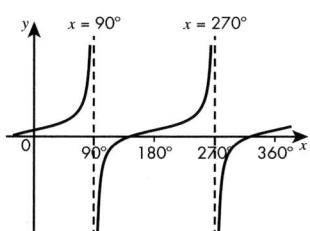

c) $y = f(x + 90°) - 1$
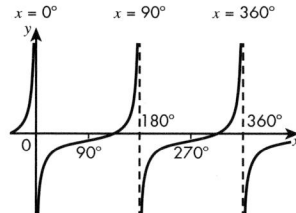

6. a) $y = -f(x + 2)$
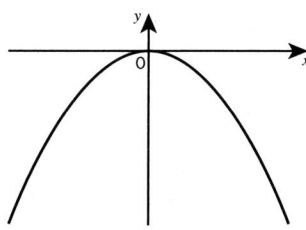

b) $y = f(x - 3) + 1$
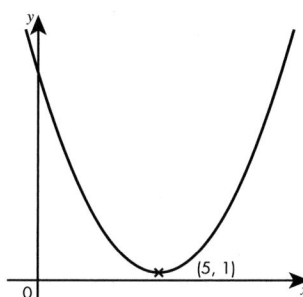

c) $y = f(-x) + 4$
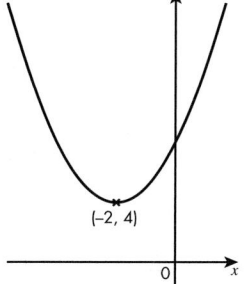

d) $y = -f(x + 1) - 3$
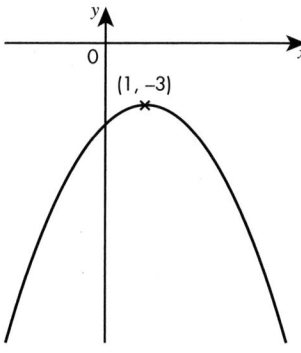

e) $y = f(-x + 1) + 2$
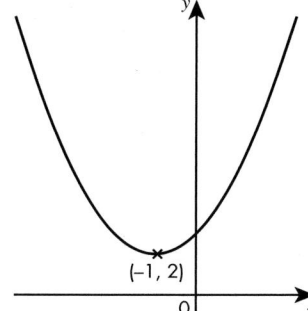

f) $y = -f(-x)$
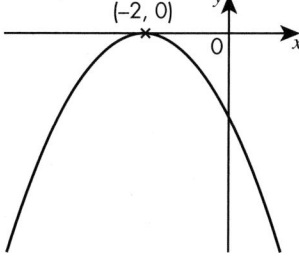

7. a) $y = -f(x + 90°)$
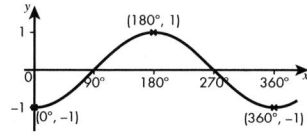

b) $y = f(-x) - 1$
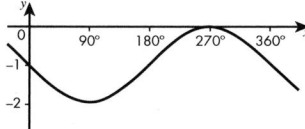

c) $y = -f(-x)$
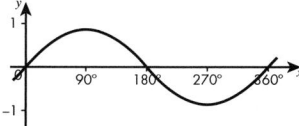

8. a) $-y = f(x) + 1$
$y = -f(x) - 1$
A$(-1, -3)$, B$(0, -4)$

b) $y - 2 = f(x) + 2$
$y = f(x) + 4$
A$(-1, 6)$, B$(0, 7)$

c) $3 - y = f(x)$
$-y = f(x) - 3$
$y = -f(x) + 3$
A$(-1, 1)$, B$(0, 0)$

4.2 Stretching graphs

Exercise 4B

1. a) $y = 2f(x)$

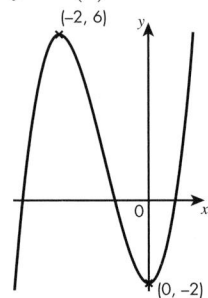

b) $y = f(2x)$

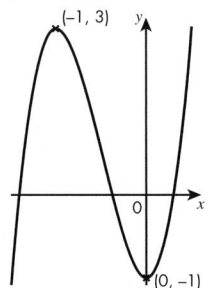

c) $y = \frac{1}{2}f(x)$

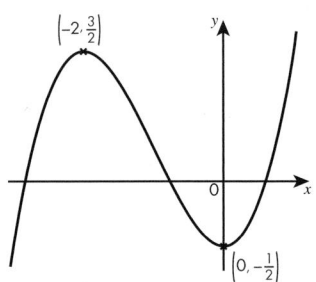

d) $y = f\left(\frac{1}{2}x\right)$

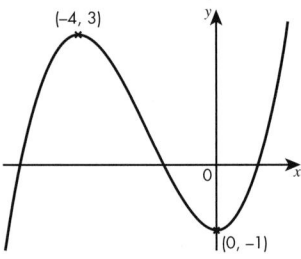

2. a) $A(-1, 6), B(0, 9), C(4, 21)$ **b)** $A(-1, -4), B(0, -6), C(4, -14)$ **c)** $A\left(-\frac{1}{5}, 2\right), B(0, 3), C\left(\frac{4}{5}, 7\right)$

d) $A(-4, 2), B(0, 3), C(16, 7)$ **e)** $A\left(\frac{1}{2}, 2\right), B(0, 3), C(-2, 7)$ **f)** $A\left(-1, \frac{2}{3}\right), B(0, 1), C\left(4, \frac{7}{3}\right)$

3. a) $y = f(x)$

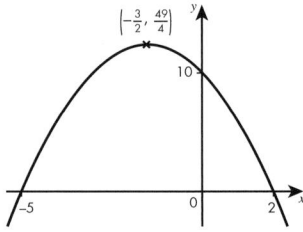

b) i) $y = 3f(x)$

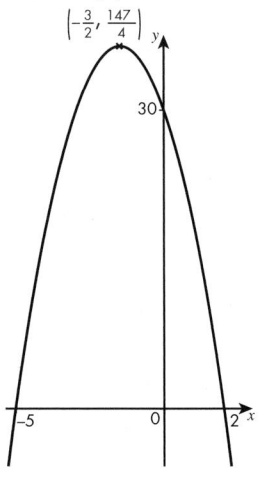

ii) $y = \frac{1}{2}f(x)$

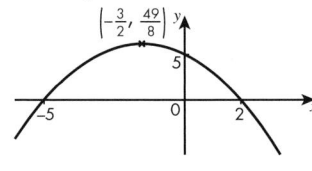

iii) $y = f(2x)$

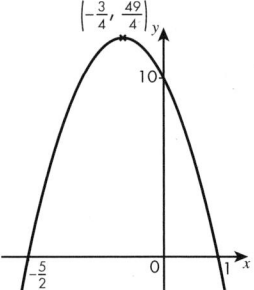

iv) $y = f\left(\frac{1}{3}x\right)$

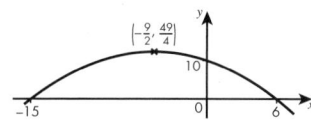

v) $y = -2f(x)$

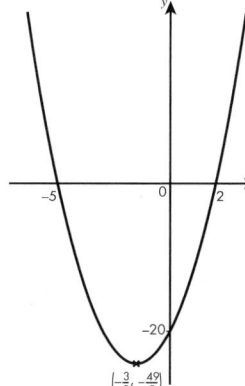

vi) $y = f(-3x)$

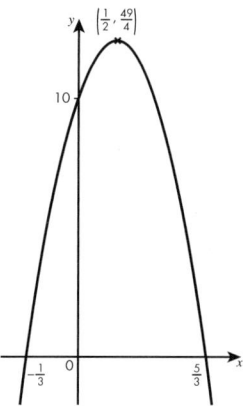

4.

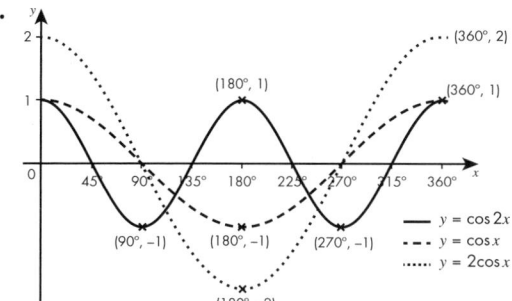

5. a) $(0, 0), (45, 3)$ and $(135, -3)$
b) $(0, 0), (90, 1)$ and $(270, -1)$
c) $(0, 0), (22.5, 2)$ and $(67.5, -2)$

6. a)

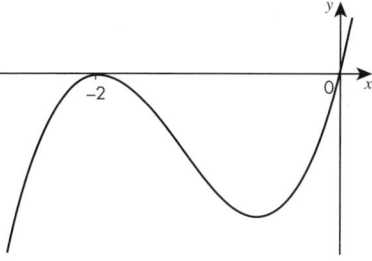

b)

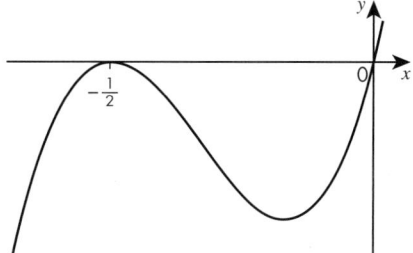

7.

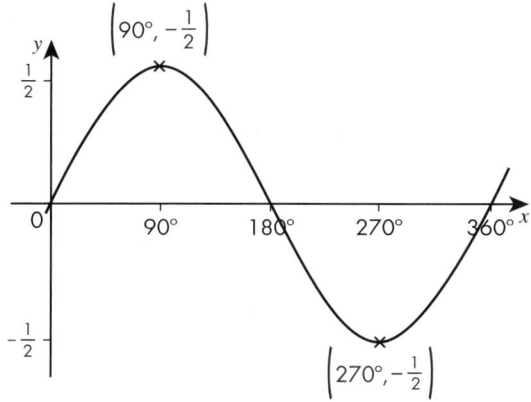

8. $P\left(\frac{2}{3}, -13\right)$

4.3 Circles

Exercise 4C

1. i) a) $x^2 + y^2 = 49$
   b)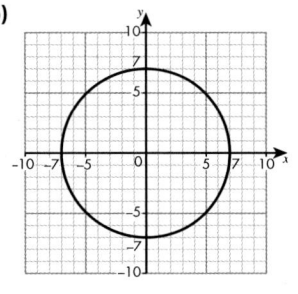

   ii) a) $x^2 + y^2 = 81$
   b)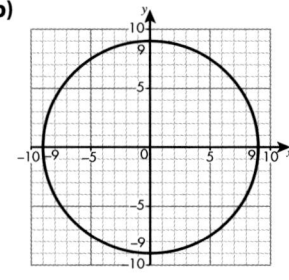

   iii) a) $(x-3)^2 + (y-3)^2 = 5$
   b)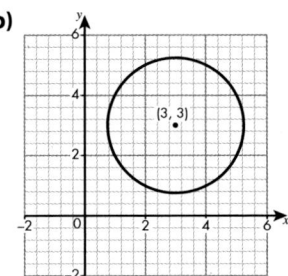

   iv) $(x-a)^2 + (y-b)^2 = 12$

2. a) 10 b) 15 c) $\sqrt{20} = 2\sqrt{5}$ d) $\sqrt{24} = 2\sqrt{6}$

3. Midpoint of AB = centre = (3, −1)
   $\frac{1}{2}$ length of AB = radius =
   $\frac{1}{2}\sqrt{(5-1)^2 + (2-(-4))^2} = \frac{1}{2} \times 2\sqrt{13} = \sqrt{13}$
   $(x-3)^2 + (y+1)^2 = 13$

4. $\begin{pmatrix} 4 \\ -7 \end{pmatrix}$

5. $(x+1)^2 + (y-2)^2 = 25$

6. a) $\frac{1}{2}$ b) −2 c) $y = -2x + 10$

7. Radius $m = -\frac{3}{8}$; Tangent $m = \frac{8}{3}$;
   $y - 3 = \frac{8}{3}(x + 8) \Rightarrow y - 3 = \frac{8}{3}x + \frac{64}{3}$
   $\Rightarrow y = \frac{8}{3}x + \frac{73}{3}$
   $\Rightarrow 3y = 8x + 73$

8. a) $y = \frac{3}{5}x - \frac{34}{5}$ b) $y = -\frac{1}{3}x - \frac{20}{3}$
   c) $y = -\frac{p}{q}x + \frac{p^2 + q^2}{q}$

9. a) 10 b) $x^2 + y^2 = 90$

10. $((-2) + 4)^2 + ((-3) - 1)^2 = r^2 = 20$; $r = \sqrt{20} = 2\sqrt{5}$

11. $4y - x = 17$

12. $y - 2x = -12$

13. $(x-3)^2 + (y+2)^2 = 10$

14. a) $(x-4)^2 + (y+3)^2 = 9$
    Centre = (4, −3)
    Radius = 3
    b) $(x+2)^2 + (y-5)^2 = 1$
    Centre = (−2, 5)
    Radius = 1
    c) $(x-3)^2 + (y-3)^2 = 32$
    Centre = (3, 3)
    Radius = $4\sqrt{2}$
    d) $(x+5)^2 + \left(y - \frac{1}{2}\right)^2 = 50$
    Centre = $\left(-5, \frac{1}{2}\right)$
    Radius = $5\sqrt{2}$

15. $(x+m)^2 - m^2 + \left(y - \frac{5}{2}n\right)^2 - \frac{25}{4}n^2 + 17 = 0$
    $\Rightarrow m = 3, \frac{5}{2}n = 4 \Rightarrow n = \frac{8}{5}$
    $(x+3)^2 + (y-4)^2 = (3)^2 + \frac{25}{4}\left(\frac{8}{5}\right)^2 - 17$
    $\Rightarrow (x+3)^2 + (y-4)^2 = 8$
    $\Rightarrow r = \sqrt{8} = 2\sqrt{2}$

4.4 Exponential and reciprocal graphs

Exercise 4D

1. a)

x	−2	−1	0	1	2	3
y	9	3	1	0.33	0.11	0.04

b) $y = \left(\frac{1}{3}\right)^x$

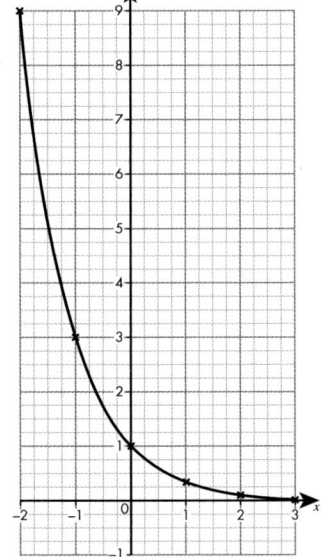

c) $x = [-0.63, -0.67]$

2. a)

x	2	3	4	5	6
y	2.5	1.67	1.25	1	0.83

b)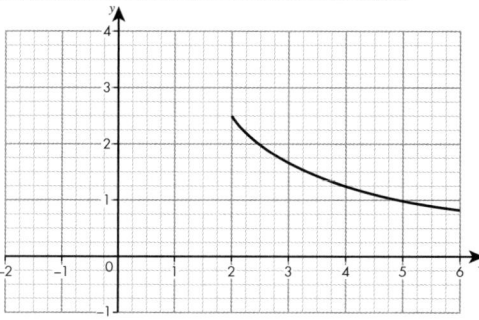

3. $3 = ab^{(0)} = a \times 1 \Rightarrow a = 3$
$81 = 3b^{(3)} \Rightarrow 27 = b^3 \Rightarrow b = 3$

4. $-2 = ab^{(0)} = a \times 1 \Rightarrow a = -2$
$-32 = -2b^{(2)} \quad 16 = b^2 \quad b = 4$
$q = -2(4)^{-2}, q = -\dfrac{1}{8}$

5. $20 = ab^2 \quad 160 = ab^5$
$a = \dfrac{20}{b^2} \qquad a = \dfrac{160}{b^5}$

$\dfrac{20}{b^2} = \dfrac{160}{b^5}$

$20b^5 = 160b^2$

$\dfrac{b^5}{b^2} = \dfrac{160}{20}$

$b^3 = 8$
$b = 2$

$a = \dfrac{20}{b^2} = \dfrac{20}{(2)^2} = \dfrac{20}{4} = 5$

6.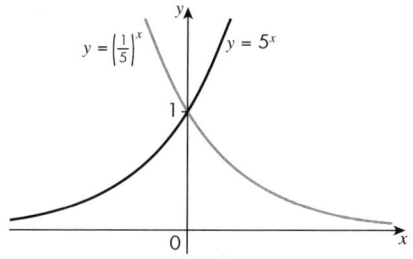

7.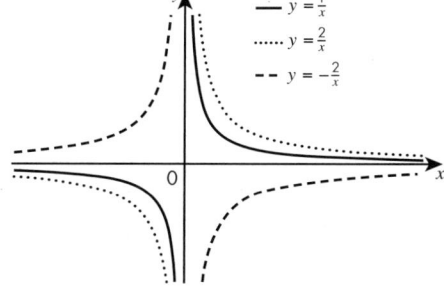

8.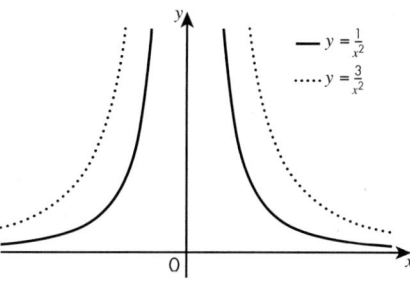

9. a) $f(x+3)$

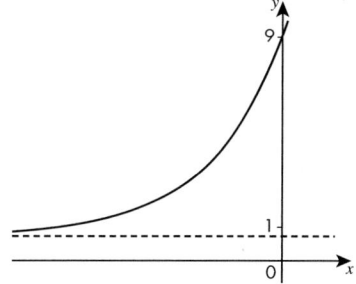

b) $f(2x)$

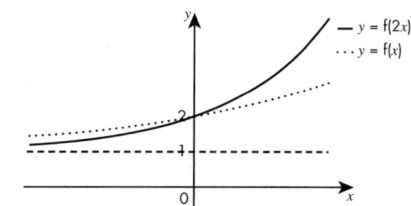

c) $-f(x)$

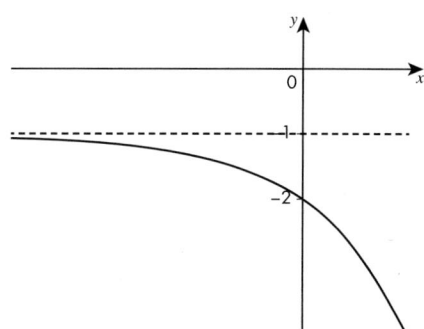

d) $f(-2x)+1$

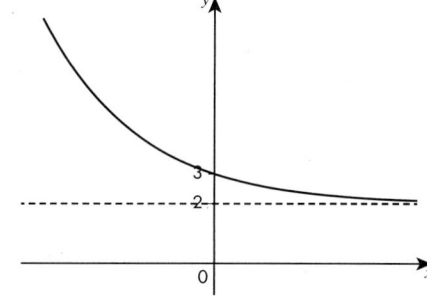

10. a) $f(x) - 3$

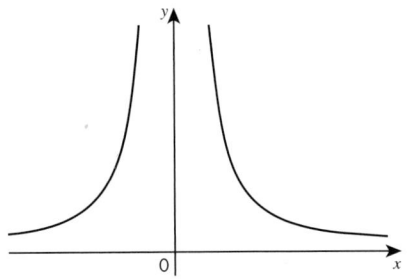

b) $2f(x)$

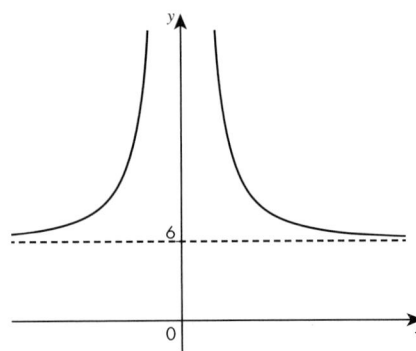

c) $f(-x)$

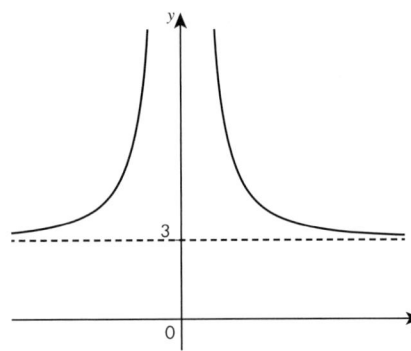

d) $-3f(x) - 1$

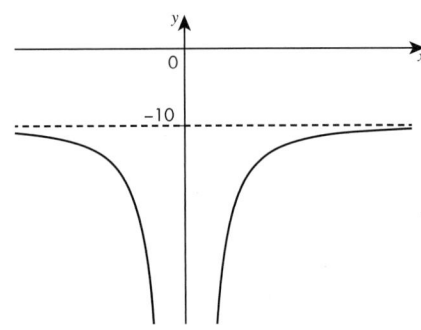

11. $y = -2f(x) - 1 = -2 \times 2^x - 1$

4.5 Non-linear graphs

Exercise 4E

1. a) The toy car is travelling at a constant speed of 4 m/s.
   b) 6 m/s = 21.6 km/h   c) 2 m/s^2
   d) 36 m   e) The gradient is steeper.
2. a) 10
   b) The number of cases never went above 30.
   c) i) $\frac{15}{40} = 0.375$ cases per day.
      ii) $\frac{8}{50} = 0.16$ cases per day.
   d) Over time, the disease is infecting fewer people.
   e) $\frac{19.5}{70} = 0.279$ cases per day.
3. a)

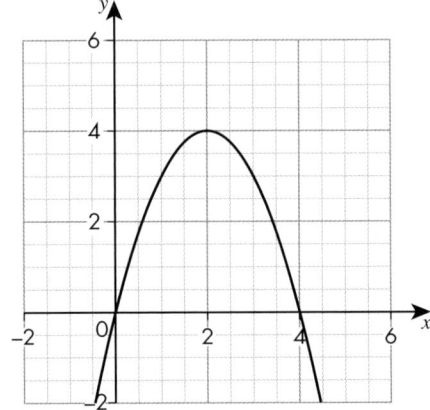

   b) $\frac{3}{1.5} = 2$ c) 10
4. a) $\frac{50}{4} = 12.5$ b) 90   c) 90
5. a) 137   b) 142.1
6. a) 18.3   b) 18.3
7. a)

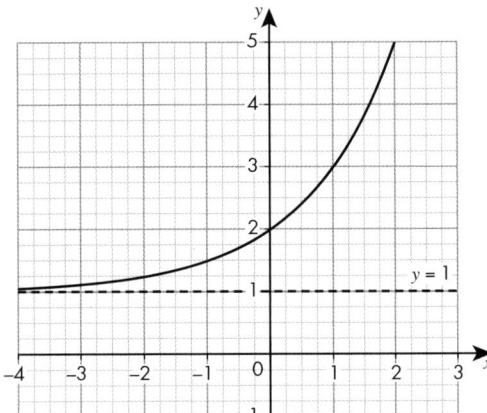

   b) 6.8

Exam style questions

1. a) $A(1, -4)$, $B\left(-\frac{7}{3}, 14\right)$ b) $A(-3, -5)$, $B\left(\frac{1}{3}, 4\right)$

2. a)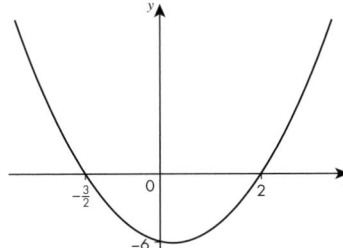

b)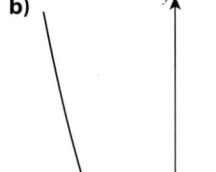

3. a) $((-1)+4)^2 + ((1)-2)^2 = 10$ b) $y = 3x + 4$
4. a) $t \approx 14$ seconds, $t = 30$ seconds.
   b) $\frac{11}{40} = 0.275$ m/s^2
   c) $\frac{10}{40} = 0.25$ m/s^2
   d) 307 m
5. a) $\left(x - \frac{1}{2}\right)^2 + (y+3)^2 = 15$
   Centre $= \left(\frac{1}{2}, -3\right)$, $r = \sqrt{15}$
   b) $\left(x - \frac{1}{2}\right)^2 + (y+3)^2 = 15$
   At y-intercept, $x = 0$.
   $\left((0) - \frac{1}{2}\right)^2 + (y+3)^2 = 15 \Rightarrow \frac{1}{4} + (y+3)^2 = 15$
   $\Rightarrow (y+3)^2 = \frac{59}{4}$
   $\Rightarrow y + 3 = \pm \frac{\sqrt{59}}{2}$
   $\Rightarrow y = -3 \pm \frac{\sqrt{59}}{2}$
   $AB = -3 + \frac{\sqrt{59}}{2} - \left(-3 - \frac{\sqrt{59}}{2}\right) = \sqrt{59}$

6.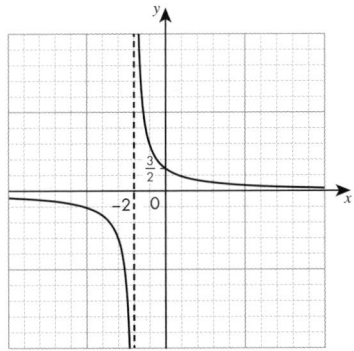

7. $(18) = ab^{(-2)}$ $\left(\frac{2}{27}\right) = ab^{(3)}$
   $18 = \frac{a}{b^2}$ $\frac{2}{27} = ab^3$
   $a = 18b^2$ $a = \frac{2}{27b^3}$
   $18b^2 = \frac{2}{27b^3}$
   $486b^5 = 2$
   $b^5 = \frac{2}{486} = \frac{1}{243}$
   $b = \frac{1}{3}$
   $a = 18\left(\frac{1}{3}\right)^2 = 18 \times \frac{1}{9} = 2$

8. $a = 3$, $b = 45$
9. 6.764

CHAPTER 5 FUNCTIONS

5.1 Functions

Exercise 5A

1. a) 3 b) $-\frac{1}{2}$
2. a) $x < 0$ b) -1 c) $x \leq -1$
   d) $-\frac{1}{2}$ e) $x = 1$ and $x = 2$
3. a) $\{10, 17, 26\}$ b) $\{1, 2, 5\}$ c) $\{y : 2 \leq y \leq 5\}$
   d) $\{y : y \geq 101\}$ e) Same as **d**
4. a) $\{0, 1, 4\}$ b) $\left\{1, \frac{1}{2}, \frac{1}{3}, \frac{1}{4}\right\}$ c) $\{5, 7, 9, 11\}$
   d) $\{5, 4, 3, 2\}$ e) $\{0, -2\}$
5. -2 can be squared so it could be in the domain. $x^2 = -2$ has no solution, so -2 cannot be in the range.
6. a) Yes b) No c) Yes
7. 5
8. a) $f(x) > 16$ b) Domain $x > 5$, range $f(x) > 61$
9. a) $f(x) > 3$ b) Domain $x > 0$, range $f(2x) > 3$
10. a) 7 b) 11
11. $a = 2$, $b = 8$

Exercise 5B

1. a) 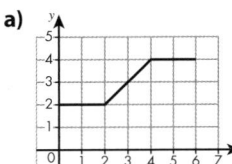 b)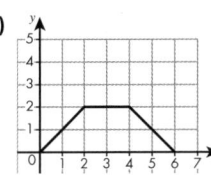

   c) 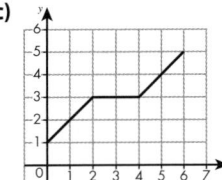 d)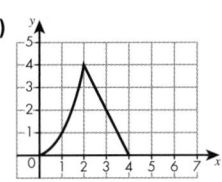

e)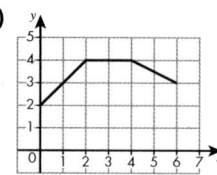
f)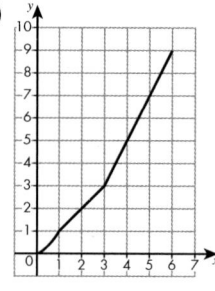

2. a) $2 \leq f(x) \leq 4$ b) $0 \leq f(x) \leq 2$ c) $1 \leq f(x) \leq 5$
   d) $0 \leq f(x) \leq 4$ e) $2 \leq f(x) \leq 4$ f) $0 \leq f(x) \leq 9$

3. a) $f(x) = 2 \quad -2 \leq x < 0$
      $= x + 2 \quad 0 \leq x < 2$
      $= 4 \quad 2 \leq x \leq 4$
   b) $f(x) = 6 \quad 0 \leq x < 2$
      $= 8 - x \quad 2 \leq x < 4$
      $= 4 \quad 4 \leq x \leq 6$
   c) $f(x) = x^2 \quad 0 \leq x < 3$
      $= 9 \quad 3 \leq x \leq 5$
   d) $f(x) = 1 \quad -4 \leq x < -1$
      $= -x \quad -1 \leq x < 1$
      $= -1 \quad 1 \leq x \leq 4$

4. Graph **a** because it represents a many-to-many mapping.

5. a)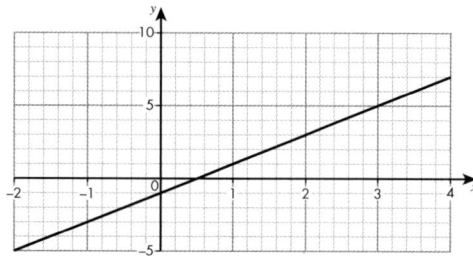
   b) $\{x \in \mathbb{R} : -2 \leq x \leq 4\}$
   c) $\{f(x) \in \mathbb{R} : -5 \leq f(x) \leq 7\}$

6. a)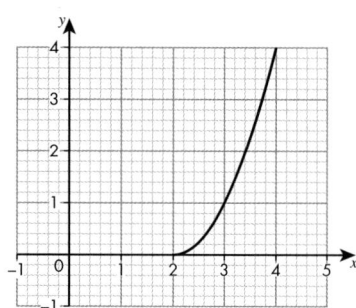
   b) $f(x) \geq 0$

7. a) $y = \dfrac{1}{x+2}$
   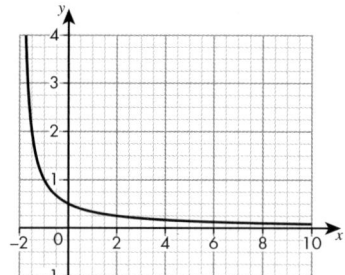
   b) $g(x) \geq 0$

8. a)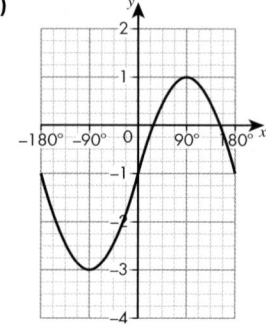
   b) $\{x \in \mathbb{R} : -180° \leq x \leq 180°\}$
   c) $\{h(x) \in \mathbb{R} : -3 \leq h(x) \leq 1\}$

5.2 Composite functions

Exercise 5C

1. a) 6 and 3   b) 7 and 3.5   c) 10 and 5
   d) $\dfrac{x+4}{2}$ e) 1 and 5   f) 1.5 and 5.5
   g) -5 and -1 h) $\dfrac{x}{2} + 4$

2. a) 1, 9, 25   b) 1, 3, 5   c) $\sqrt{2x+1}$

3. a) 6 and 18   b) 12 and 36   c) $9x$

4. a) $3(x-6)$ b) $3x - 6$

5. $ab(x) = x - 7 + 4 = x - 3$; $ba(x) = x + 4 - 7 = x - 3$

6. a) 7   b) 8   c) 256   d) 21

7. a) $6x$ b) $6x - 5$

8. a) $9x^2 + 24x + 16$ or $(3x+4)^2$ b) $6x - 5$
   c) $2x + 3$ d) $4 - 2x$

9. a) $x - 10$ b) $x + 10$ c) x

10. a) x^4 b) $\left(\dfrac{12}{x}\right)^2$ or $\dfrac{144}{x^2}$ c) $\dfrac{12}{x^2}$ d) x

11. a) 80   b) $2(2x - 1) - 1 = 4x - 3$
    c) $(2x-1)^2 + 2(2x-1) = 4x^2 - 4x + 1 + 4x - 2 = 4x^2 - 1$

12. a) $\dfrac{1}{3x-3}$ b) $\dfrac{x-1}{x-4}$

13. $x = 0$ or $x = -\dfrac{1}{2}$

14. $x = 1$ or 3

15. $x = 5$

5.3 Inverse functions

Exercise 5D

1. **a)** $x - 7$ **b)** $\frac{x}{8}$ **c)** $5x$ **d)** $x + 3$
2. **a)** $3(x + 2)$ **b)** $\frac{x}{4} + 5$ **c)** $5x - 4$ **d)** $\frac{2x + 6}{3}$
   e) $2\left(\frac{x}{3} - 4\right)$ **f)** $\sqrt[3]{\frac{x}{4}}$
3. **a)** $\frac{8}{x}$ **b)** $\frac{20}{x+1}$ **c)** $\frac{2}{x} - 1$
4. **a)** $\sqrt{x - 4} + 2$ **b)** $\frac{4x - 1}{3}$ **c)** $\frac{8}{5x + 3}$
   d) $\frac{9x}{5} + 32$ **e)** $\sqrt[3]{x} + 2$
5. $f(x) = f^{-1}(x)$
6. $g(x)$ tends towards infinity
7. $h^{-1}(x) = x^3 - 2$

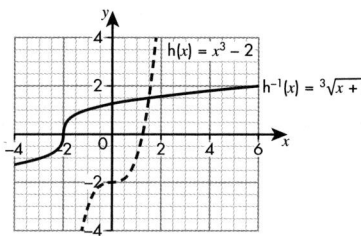

8. **a)** The range is $\{g(x) \in \mathbb{R} : g(x) > 0\}$,
   b) $g^{-1}(x) = x^2 + 6$. The domain is $\{x \in \mathbb{R} : x > 0\}$, the range is $\{g^{-1}(x) \in \mathbb{R} : g^{-1}(x) > 6\}$,
   c)

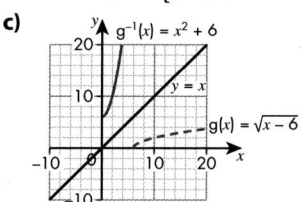

   d) A reflection in the line $y = x$
9. $f^{-1}(x) = \frac{(3x + 5)^2}{2}$, domain $\left\{x \in \mathbb{R}: x \geqslant -\frac{5}{3}\right\}$, range $\{f^{-1}(x) \in \mathbb{R}: f^{-1}(x) \geqslant 0\}$
10. $f^{-1}(x) = \frac{5x - 3}{x - 4}$, domain $\{x \in \mathbb{R}: x > 4\}$, range $\{f^{-1}(x) \in \mathbb{R}: f^{-1}(x) > 5\}$

5.4 Transforming functions

Exercise 5E

1. **a)** $9x - 6$ **b)** $3x + 5$ **c)** $6x - 2$ **d)** $3x - 8$
2. Stretch in the x-direction scale factor $\frac{1}{3}$
3. Reflection in the y-axis ($x = 0$)
4. **a)** $f(x) = -x(x + 4)$ **b)** $g(x) = f(x) + 2$
   c) $h(x) = -f(x) - 1$
5. $y = f(x - 4) - 1$
6. $y = 2g\left(\frac{1}{3}x\right)$
7. $y = \frac{(-3x)^2 + 1}{7} - 2 = \frac{(-3x)^2 - 13}{7} = \frac{9x^2 - 13}{7}$
8. **a)** $y = \sin x$ **b)** $y = -2\sin x + 3$

Exam-style questions

1. 110 and 90
2. **a) i)** one-to-one **ii)** many-to-one
   iii) one-to-many **iv)** many-to one
   b) iii) is not a function
3. **a)** $x = 19.25$ **b)** $\{f(x) \in \mathbb{R} : f(x) \geqslant 15\}$,
4. **a)** -4 **b)** $g^{-1}(x) = \frac{2 - 4x}{x - 1}, x \neq 1$
5.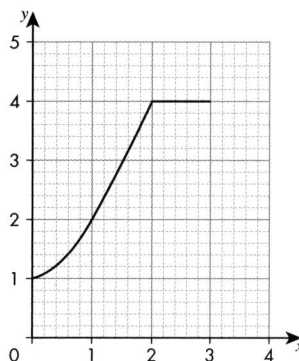
6. **a)** translation $\begin{pmatrix} 1 \\ 0 \end{pmatrix}$, stretch in the y-direction scale factor 3, translation $\begin{pmatrix} 0 \\ -2 \end{pmatrix}$
   b) stretch in the x-direction scale factor 2, reflection in the x-axis, translation $\begin{pmatrix} 0 \\ 3 \end{pmatrix}$
7. $g(x) = -(x + 1)^2 + 2$
8. **a)**

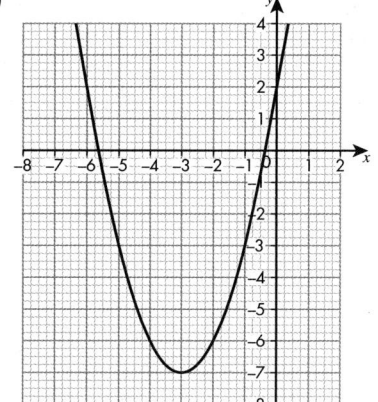

   Range $\{f(x) \in \mathbb{R}: f(x) \geqslant -7\}$

   b) $f(x)$ is a many-to-one function and therefore does not have an inverse
9. $ff(x) = \frac{x - 2}{-2x + 5}$

CHAPTER 6 EQUATIONS AND INEQUALITIES

6.1 Solve equations

Exercise 6A

1. **a)** $x = 7$ **b)** $x = 5$ **c)** $x = -2$
2. **a)** $x = 3$ **b)** $x = 6$ **c)** $x = 2$
3. **a)** $x = 6$ **b)** $x = 14$ **c)** $x = 7$
4. 5, 6 and 7
5. 50, 55 and 75 degrees
6. **a)** $x = -1$ **b)** $x = \frac{1}{2}$ **c)** $x = -\frac{25}{6}$
   d) $x = -\frac{13}{4}$ **e)** $x = \frac{9}{7}$ **f)** $x = -\frac{7}{2}$
7. **a)** $x = 60°, 120°$ **b)** $x = 45°, 225°$
   c) $x = 45°, 315°$ **d)** $x = 60°, 300°$
8. **a)** $x = -150°, 90°$ **b)** $x = -15°, -75°$
   c) $x = -160°, -100°, -40°, 20°, 80°, 140°$
   d) $x = -157.5°, -112.5°, 22.5°, 67.5°$

6.2 Solve quadratic equations

Exercise 6B

1. **a)** $\frac{1}{3}, -3$ **b)** $1\frac{1}{3}, -\frac{1}{2}$ **c)** $-2\frac{1}{2}, 3\frac{1}{2}$
   d) $-\frac{1}{6}, -\frac{1}{3}$ **e)** $\frac{5}{2}, -\frac{7}{6}$ **f)** $\frac{7}{5}, -\frac{5}{3}$
   g) $1\frac{3}{4}, 1\frac{2}{7}$ **h)** $\pm \frac{1}{4}$ **i)** $-2\frac{1}{4}, 0$ **j)** $-\frac{1}{3}, 3$
2. **a)** $-3, 4$ **b)** $-\frac{5}{2}, \frac{3}{2}$ **c)** $-6, 7$ **d)** $-1, \frac{11}{13}$
   e) $-\frac{1}{2}, -\frac{1}{3}$ **f)** $-2, \frac{1}{5}$ **g)** 4 **h)** $-2, \frac{1}{8}$
3. **a)** Both have only one solution: $x = 1$.
   b) B is a linear equation, but A and C are quadratic equations.
4. **a)** $(5x - 1)^2 = (2x + 3)^2 + (x + 1)^2$, when expanded and collected into the general quadratic form, gives the required equation.
   b) $(10x + 3)(2x - 3)$, $x = 1.5$; area = 7.5 cm².
5. **a)** $2(x + 2) + 5(x + 1) = 3(x + 1)(x + 2)$ which simplifies to the required equation.
   b) $4(x + 1) + 7(x - 2) = 3(x - 2)(x + 1)$ which simplifies to the required equation.
   c) $3(3x - 1) - 4(2x - 1) = (2x - 1)(3x - 1)$ which simplifies to the required equation.
6. **a)** 3, −1.5 **b)** 4, −1.25 **c)** 3, −2.5
7. **a)** $x = 2$ and $x = 5$ **b)** $x = 4$ and $x = 25$

Exercise 6C

1. 1.77, −2.27
2. −0.23, −1.43
3. 3.70, −2.70
4. 1.64, 0.61
5. 0.36, −0.79
6. 1.89, 0.11
7. 13
8. $x^2 - 3x - 7 = 0$
9. $b^2 - 4ac < 0$ or $b^2 - 4ac = -8$ so no real solution as a negative number does not have a real square root.
10. **a)** 1 **b)** 0 **c)** 2

Exercise 6D

1. **a)** $-2 \pm \sqrt{5}$ **b)** $-7 \pm 3\sqrt{6}$ **c)** $3 \pm \sqrt{6}$
   d) $-1 \pm \sqrt{2}$ **e)** $1 \pm 2\sqrt{2}$ **f)** $-1 \pm \sqrt{10}$
2. **a)** 1.45, −3.45 **b)** 5.32, −1.32 **c)** −4.16, 2.16
3. third, first, fourth, second, in that order.
4. **a)** $x = 1.5 \pm \sqrt{3.75}$ **b)** $x = 1 \pm \sqrt{0.75}$
   c) $x = -1.25 \pm \sqrt{6.5625}$ **d)** $x = 7.5 \pm \sqrt{40.25}$
5. **a)** $-1 \pm \sqrt{10}$ **b)** $\frac{-3 \pm \sqrt{15}}{2}$ **c)** $\frac{-4 \pm \sqrt{46}}{3}$
   d) $\frac{-5 \pm \sqrt{73}}{4}$
6. $\left(x + \frac{5}{4}\right)^2 = -\frac{23}{16}$; we cannot square root a negative to get a real number; therefore there are no real roots.
7. **a)** $\frac{57}{8} - 2\left(x - \frac{3}{4}\right)^2$ **b)** $x = \frac{3 \pm \sqrt{57}}{4}$
   c)

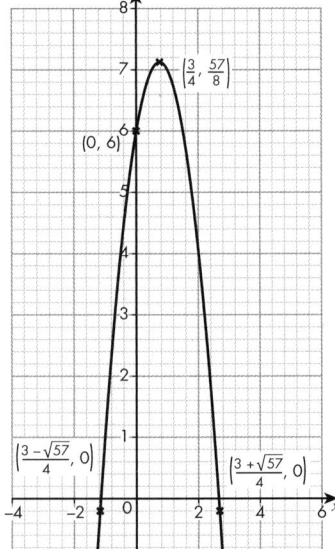

$y = -2x^2 + 3x + 6$

6.3 Solve simultaneous equations

Exercise 6E

1. **a)** (5, −1) **b)** (4, 1) **c)** (8, −1)
2. **a)** (1, 2) and (−2, −1) **b)** (−4, 1) and (−2, 2)
3. **a)** (3, 4) and (4, 3) **b)** (0, 3) and (−3, 0)
   c) (3, 2) and (−2, 3)

4. a) (2, 5) and (−2, −3)   b) (−1, −2) and (4, 3)
   c) (3, 3) and (1, −1)
5. a) (−3, −3), (1, 1)   b) (3, −2), (−2, 3)
   c) (−2, −1), (1, 2)   d) (2, −1), (3, 1)
6. a) (2, 3) $\left(-\frac{2}{3}, \frac{1}{3}\right)$
   b)
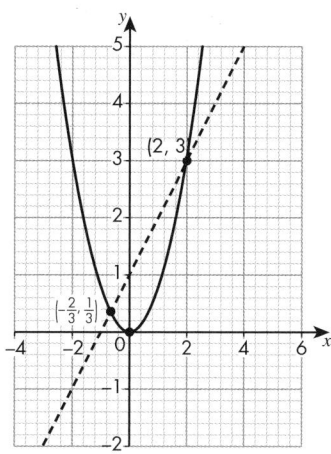
7. a) (−2, −2), $\left(\frac{2}{5}, \frac{14}{5}\right)$
   b)
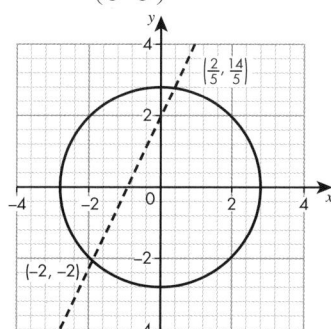
8. a) $(5 - 2\sqrt{2}, \sqrt{2} - 4)$ $(5 + 2\sqrt{2}, -\sqrt{2} - 4)$
   b) $\left(\frac{2 - \sqrt{42}}{2}, -\sqrt{\frac{21}{2}}\right) \left(\frac{2 + \sqrt{42}}{2}, \sqrt{\frac{21}{2}}\right)$

Exercise 6F
1. a) $\frac{1}{2}$ and $\frac{7}{2}$ b) 3 and 4   c) 7 and 2
2. No as the resulting quadratic $3x^2 − 10x + 18 = 0$ has no real solutions $\left(\text{after substituting } y = \frac{5 - 3x}{3}\right)$
3. $x = 7.5$ and $y = 3$
4. a) i) $x^2 + y^2 = 36$ ii) $x + y = 3$
   b) $\left(\frac{3(1-\sqrt{7})}{2}, \frac{3(1+\sqrt{7})}{2}\right), \left(\frac{3(1+\sqrt{7})}{2}, \frac{3(1-\sqrt{7})}{2}\right)$

6.4 Solving inequalities

Exercise 6G
1. a) $x \geq -6$ b) $y \leq 4$ c) $x \leq \frac{14}{5}$
2. a) $x \geq 7.5$ b) $x \leq -2$ c) $x < 6$ d) $x > 1.5$
   e) $x \geq -5$ f) $x < 0.5$

3. a) $3 < x < 6$ b) $2 < x < 5$ c) $-1 \leq x \leq 3$
   d) $1 \leq x < 4$ e) $2 \leq x < 4$ f) $0 \leq x \leq 5$

Exercise 6H
1. a) $-4 \leq x \leq 4$ b) $-2 < x < 2$
   c) $x < -2.5$ or $x > 2.5$ d) $x \leq -1$ or $x \geq 1$
2. a) $-3 < x < 3$ b) $x < -5$ or $x > 5$
   c) $x \leq -1.5$ or $x \geq 1.5$ d) $-0.5 \leq x \leq 0.5$
3. a) $-2 \leq x \leq 2$ b) $x < -3.5$ or $x > 3.5$
   c) $-2.5 < x < 2.5$ d) $-3 \leq x \leq 3$
4. a) $-3 \leq x \leq 1$ b) $x > \frac{2}{3}$ or $x < -3$
   c) $-1 < x < \frac{5}{9}$ d) $-5 \leq x \leq \frac{3}{5}$ e) $-\frac{1}{3} \leq x \leq 4$
5. 2, 3, 4
6. 1, 2
7. a) $x < -2, x > 5$ b) $-7 < x < -5$
   c) $-8 \leq x \leq 9$ d) $\frac{1}{3} \leq x \leq 3$
   e) $x < -\frac{11}{2}, x > -1$ f) $x \leq \frac{3}{5}, x \geq 2$

Exercise 6I
1. a)
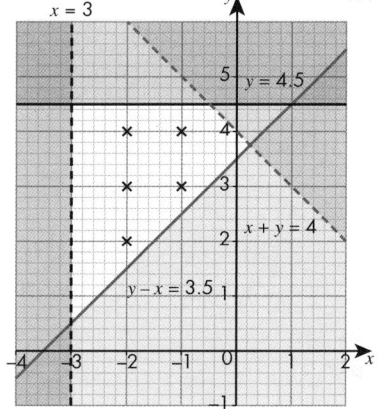
   b) (−2, 2), (−2, 3), (−2, 4), (−1, 3) and (−1, 4)
2. $x < 2, y > 1, x + y \geq 2, 2x + 3y \leq 8$
3. a) $x \geq 10, y \geq 20, 2x + y \leq 60, x + y > 40$
   b)
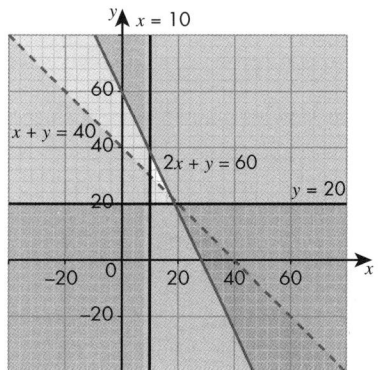
   c) This indicates the unshaded region.
   d) 10 pens and 40 pencils costing £15

4. **a)** $x \geqslant 25, y < 60$

   b)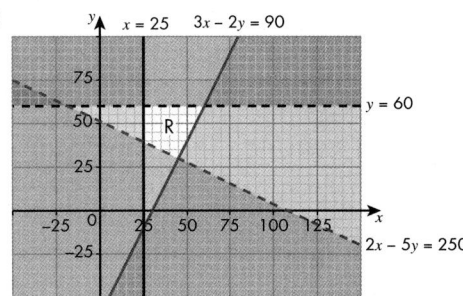

   c) Cost (C) $= x + 2y$

   d) 50 medals and 30 trophies

5.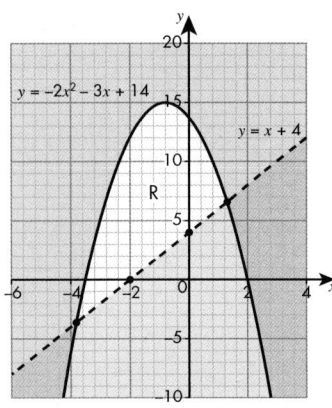

Exam-style questions

1. $x = 9$
2. $x = \frac{9}{4}$
3. $x = 30°, 150°$
4. width = 2.46 m and length = 4.46 m
5. (3, 0) and (8, −5)
6. **a)** $a = 3, b = -2, c = -8$ **b)** $x = 3.63$ or 0.37
7. **a)** $x(3x + 4) \geqslant 160$
   b) $x \leqslant -8$ or $x \geqslant 6\frac{2}{3}$; AD is at least $6\frac{2}{3}$ cm
8. $x < 2$
9. −1, 0, 1, 2, 3
10. $x < 2, x > 10$
11. **a)**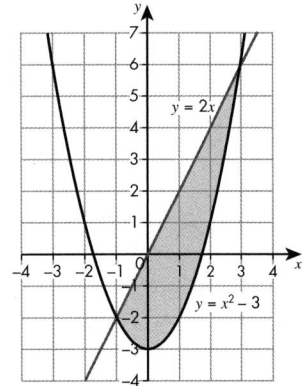

    b) $-1 \leqslant x \leqslant 3$

CHAPTER 7 PYTHAGORAS' THEOREM AND TRIGONOMETRY

7.1 Pythagoras' theorem in 2D and 3D

Exercise 7A

1. **a)** 32.2 cm² **b)** 2.83 cm² **c)** 50.0 cm²
2. **a)** **i)** 14.4 cm **ii)** 13 cm **iii)** 9.4 cm
   b) 15.3 cm (1 dp)
3. No, 6.55 m is longest length
4. **a)** 20.6 cm **b)** 15.0 cm
5. 21.3 cm
6. $\sqrt{300} = 17.3$ cm
7. 20.6 cm
8. **a)** 50.0 cm **b)** 54.8 cm **c)** 48.3 cm
   d) 27.0 cm

7.2 Trigonometry in 2D and 3D

Exercise 7B

1. 65°
2. The safe limits are between 1.04 m and 2.05 m. The ladder will reach between 5.64 m and 5.91 m up the wall.
3. 31°
4. **a)** 338 km **b)** 725 km
5. 43.5 km

Exercise 7C

1. 10.1 km
2. 22°
3. **a)** 156 m
   b) No. The new angle of depression is $\tan^{-1}\frac{200}{312} = 33°$ and half of 52° is 26°.
4. **a)** 222 m **b)** 42°
5. The angle is 16° so Cara is not quite correct.

Exercise 7D

1. 25.1°
2. **a)** 25 cm **b)** 58.6° **c)** 20.5 cm
3. **a)** 3.46 m **b)** 75.5° **c)** 73.2°
4. **a)** It is 44.6°; use triangle XDM where M is the midpoint of BD; triangle DXB is isosceles, as X is over the point where the diagonals of the base cross; the length of DB is $\sqrt{656}$ and the cosine of the required angle is $0.5\sqrt{656} \div 18$.
   b) 57.7°

7.3 Sine and cosine rule and area of a triangle in 2D and 3D

Exercise 7E
1. a) 3.64 m   b) 8.05 cm   c) 19.4 cm
2. a) 46.6°   b) 112.0°   c) 36.2°
3. 3.47 m
4. 22.2 m
5. 64.6 km
6. a) $\sqrt{208} = 14.4$ cm   b) $\sqrt{288} = 17.0$ cm
   c) 64°

Exercise 7F
1. a) 7.71 m   b) 29.1 cm   c) 27.4 cm
2. a) i) 76.2°   ii) 125.1°   iii) 90°
   b) Right-angled triangle
3. a) 10.7 cm   b) 41.7°   c) 38.3°   d) 6.7 cm
4. 58.4 km at 092.5°
5. 21.8°; the smallest angle is opposite the shortest side.
6. 111°; the largest angle is opposite the longest side.
7. 56.4°

Exercise 7G
1. a) 8.6 m   b) 90°   c) 27.2 cm   d) 26.9°
   e) 27.5°   f) 62.4 cm
2. 7 cm
3. 11.1 km
4. a) $A = 90°$; this is Pythagoras' theorem
   b) A is acute   c) A is obtuse
5. 142 m or 143 m
6. 221°
7. $\sqrt{2}$ cm, 90°, 45°, 45°
8. 8 cm^2

Exercise 7H
1. a) 3 cm   b) $4\sqrt{3}$ cm   c) 1.5 cm
   d) $\frac{5}{\sqrt{3}}$ cm or $\frac{5\sqrt{3}}{3}$ cm   e) $3\sqrt{2}$ cm   f) 2 cm
2. a) $3\sqrt{2}$ cm   b) $(5\sqrt{3} - 5)$ cm
   c) $\frac{3\sqrt{2}}{4}$ cm

Exam-style questions
1. 84 cm^2
2. 8 cm
3. $\frac{7}{32}$
4. 5.16 cm
5. 17.3 cm

6. a) 12.3 cm   b) 29°   c) 59°
7. a) In triangle VMD: VM^2 = 9^2 − 4^2, so VM = $\sqrt{65}$
   b) 63°
8. 193 cm^2
9. 532 cm^2
10. $16 + 4\sqrt{7}$

CHAPTER 8 PROBABILITY

8.1 The language of probability

Exercise 8A
1. a) $\frac{16}{100}$ b) $\frac{36}{100}$ c) $\frac{48}{100}$ d) $\frac{84}{100}$
   e) 0   f) Yes, the probabilities of drawing the second counter would be different.

Exercise 8B
1. 4 or 9 bananas
2. a) 0.48   b) 0.6
3. $0.05 \times 0.15 \neq 0.05$ so not independent
4. $0.1 \times 0.3 = 0.03$ so independent
5. $\frac{2}{5} \times \frac{1}{5} = \frac{2}{25}$ so independent
6. $\frac{1}{4} \times \frac{3}{10} = \frac{3}{40}$ so independent
7. 0.6
8. $\frac{1}{5}$
9. 0.25
10. a)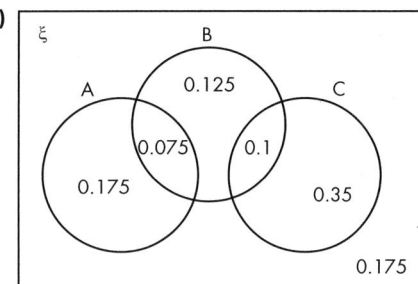

   b) i) 0.075   ii) 0   iii) 0.175   iv) 0.3   v) 0.625
11. a) $\frac{1}{12}$
   b)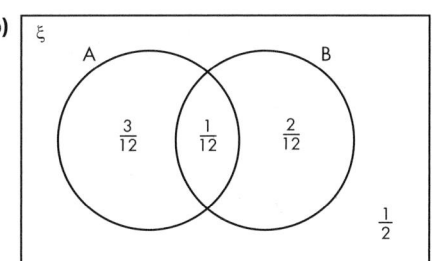

   c) $\frac{1}{2}$

8.2 Conditional probability

Exercise 8C

1. a)

	Circle	Square	Triangle	Total
Red	39	45	47	131
Blue	34	68	17	119
Total	73	113	64	250

b) $\frac{34}{250}$ **c)** $\frac{45}{113}$

2. a)

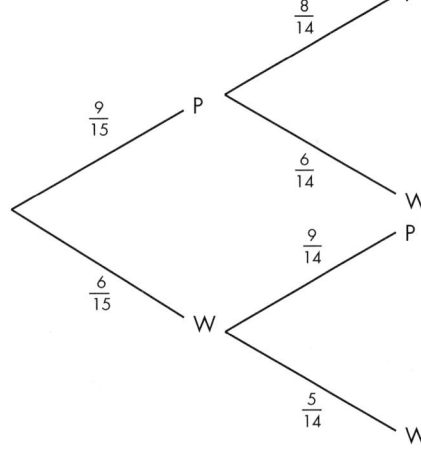

b) i) $\frac{5}{35}$ **ii)** $\frac{12}{35}$ **iii)** $\frac{18}{35}$ **iv)** $\frac{23}{35}$

3. a)

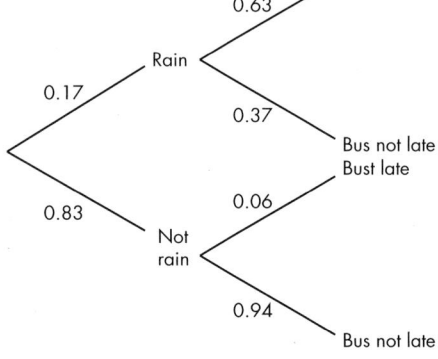

b) 0.0498

c) 0.8431

4. $\frac{154}{435}$

Exercise 8D

1. a) $\frac{9}{100}$ **b)** $\frac{11}{20}$ **c)** $\frac{3}{13}$ **d)** $\frac{10}{13}$ **e)** $\frac{9}{20}$

2. a) $3x^2 - 28x + 60 = 0$ **b)** $x = 6$

3. 16

4. $x = 3, y = 12$

5. a)

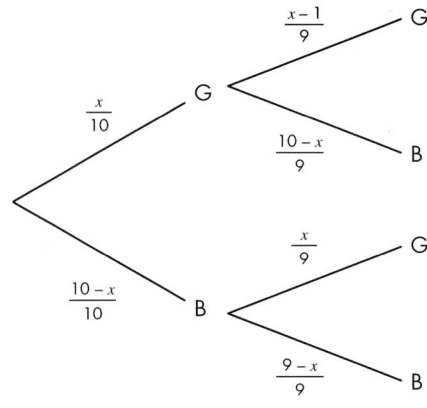

b) $\frac{x(10-x)}{45}$

c) $x = 7$

Exam-style questions

1. a)

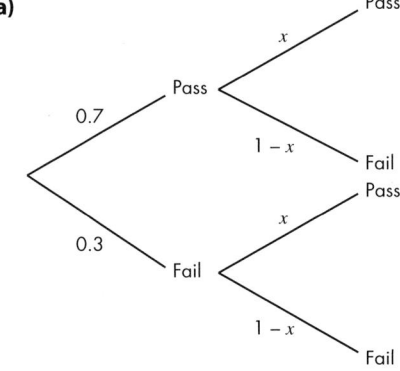

b) Independent events

c) $x = 0.45$

2. 0.615

3. $P(T,T,T) = 0.67 \times 0.67 \times 0.67 = 0.300763 > 0.3$, so Abdi is wrong.

4. a) $x = \frac{1}{6}, y = \frac{1}{15}, z = \frac{1}{4}$

   b) $P(A) \times P(B) = P(A \cap B)$ $\frac{5}{12} \times \frac{2}{5} = \frac{1}{6}$

5. $2x(1-x) = 0.455$

$x^2 - x - 0.2275 = 0$

$x = 0.65$, so the probability that he loses the next game is 0.35.

CHAPTER 9 PROOF

9.1 Proof by deduction

Exercise 9A

1. a) Odd, yes

   b) $(2n+1) + 2m = 2(n+m) + 1$, which is odd.

2. **a)** For example, $2m + 2n = 2(m + n)$, which is even.
   b) For example, $(2m)(2n) = 2(2mn)$, which is even.
   c) For example, $(2m + 1)(2n) = 2(2mn + n)$, which is even.
   d) For example, $(2m + 1)(2n + 1) = 2(2mn + m + n) + 1$, which is odd.
   e) For example, $n + (n + 1) + (n + 2) + (n + 3) = 2(2n + 3)$, which is even.
   f) For example, from **e**, $\frac{1}{2} \times 2(2n + 3) = 2(n + 1) + 1$, which is odd.

3. **a)** 3, 5, 8, 13, 21, 34, 55
   b) $3a + 5b, 5a + 8b, 8a + 13b, 13a + 21b, 21a + 34b$
   c) $(8a + 13b) - (2a + 3b) = 6a + 10b = 2(3a + 5b)$

4. **a)** Substitute $n = 11$ and $n = 12$ into the formula to give 66 and 78 and add them to get 144, which is a square number.
   b) Substitute $(n + 1)$ for n in the formula:
   $\frac{1}{2}(n + 1)(n + 2) + 1 = \frac{1}{2}n^2 + 3n + 2 = \frac{1}{2}(n + 1)(n + 2)$
   c) $\frac{1}{2}n(n + 1) + \frac{1}{2}(n + 1)(n + 2)$, which simplifies to $(n + 1)^2$, which is a square number.

5. **a)** $\frac{1}{2}(6)(6 + 1) = 21 = 1 + 2 + 3 + 4 + 5 + 6$
   b) The sum of each respective pair of terms of the two series is $(n + 1)$, so the sum is $n(n + 1)$
   c) From **b**, twice the sum of the first n integers is $n(n + 1)$.
   Hence the sum is $\frac{1}{2}n(n + 1)$

6. **a)** $8T + 1 = 8 \times \frac{1}{2}n(n + 1) + 1$ which simplifies to $(2n + 1)^2$, which is a square number.
   b) $9T + 1 = 9 \times \frac{1}{2}n(n + 1) + 1 = \frac{1}{2}(9n^2 + 9n + 2)$
   $= \frac{1}{2}(3n + 1)(3n + 2)$
   $= \frac{1}{2}(3n + 1)((3n + 1) + 1)$, which is the $(3n + 1)$th triangular number

7. **a)** Use the fact that any two numbers can be written as $(a + b)$ and $(a - b)$. So, let $x = (a + b)$ and $y = (a - b)$.
   Then $\frac{1}{2}[(a + b)^2 + (a - b)^2]$ can be expanded and simplified to give $a^2 + b^2$, which is the sum of two squares.
   b) Let $x = (a + b)$ and $y = (a - b)$.
   Then $2[(a + b)^2 + (a - b)^2]$ can be expanded and simplified to $(2a)^2 + (2b)^2$, which is the sum of two squares.
   c) This can be shown to be true.

8. LHS $= (2n)^2 + (n^2 - 1)^2$
   $= 4n^2 + n^4 - 2n^2 + 1$
   $= n^4 + 2n^2 + 1$
   $= (n^2 + 1)^2$
   $=$ RHS

9. Yes, this has been shown to be true.

10. $(2x + 1)^2 - (x - 1)^2 = 4x^2 + 4x + 1 - (x^2 - 2x + 1)$
    $= 4x^2 + 4x + 1 - x^2 + 2x - 1$
    $= 3x^2 + 6x = 3x(x + 2)$
    $= (2x + 1 + x - 1)(2x + 1 - (x - 1))$

11. $\frac{\sqrt{12} + 5}{\sqrt{3} + 2} = \frac{2\sqrt{3} + 5}{\sqrt{3} + 2} = \frac{2\sqrt{3} + 5}{\sqrt{3} + 2} \times \frac{\sqrt{3} - 2}{\sqrt{3} - 2} =$
    $\frac{6 - 4\sqrt{3} + 5\sqrt{3} - 10}{3 - 4} = \frac{\sqrt{3} - 4}{-1} = 4 - \sqrt{3}$

12. $\frac{1}{\frac{1}{\sqrt{5}} + \sqrt{5}} = \frac{1}{\frac{1 + 5}{\sqrt{5}}} = \frac{\sqrt{5}}{6}$

Exercise 9B

1. $\frac{1}{3}(x - 3)^2 + 2$; minimum turning point is (3, 2) and the range of $\frac{x^2}{3} - 2x + 5$ is $y \geq 2$ (and so positive for all values of x)

2. $-2\left(x + \frac{1}{4}\right)^2 - \frac{23}{8}$; maximum turning point is $\left(-\frac{1}{4}, -\frac{23}{8}\right)$ and the range of $-2x^2 + x - 3$ is $y \leq -\frac{23}{8}$ (and so negative for all values of x)

9.2 Proof by exhaustion and disproof by counter example

Exercise 9C

1. $1 \times 3 = 3, 3 \times 5 = 15, 5 \times 7 = 35$, all results are odd numbers, so statement is proved by exhaustion.

2. $17 \div 2 = \frac{17}{2}, 17 \div 3 = \frac{17}{3}, 17 \div 4 = \frac{17}{4}$, as $5^2 > 17$ can stop there as all possible factors have been tested. There are no integer factors other than 1 and 17, so 17 is prime.

3. The six possible arrangements are: TOOT, TOTO, TTOO, OOTT, OTOT, OTTO. There are no others, so the statement is proved by exhaustion.

4. Any negative number e.g. $x = -2, (-2)^2 = 4, \sqrt{4} = \pm 2$, not just -2.

5. $3^2 - 1^2 = 9 - 1 = 8$, which is even.

Exercise 9D

1. $(2n)^3 + 3(2n) + 1 \Rightarrow 2(4n^3 + 3n) + 1$ (always odd) and $(2m + 1)^3 + 3(2m + 1) + 1 \Rightarrow 2(4m^3 + 6m^2 + 6m + 2) + 1$ (always odd)

2. $n = 2 \Rightarrow 2^2 - 1 = 3$ (prime), $n = 3 \Rightarrow 2^3 - 1 = 7$ (prime)

3. $n = 4 \Rightarrow 2^4 - 1 = 15 = 3 \times 5$ (not prime)

4. $n = 1 \Rightarrow 1^2 + 1 = 2, n = 2 \Rightarrow 2^2 + 2 = 6, n = 3 \Rightarrow 3^2 + 3 = 12, n = 4 \Rightarrow 4^2 + 4 = 20$; all divisible by 2

5. $\theta = 30° \Rightarrow \sin(2 \times 30°) = 60° = \frac{\sqrt{3}}{2}, 2\sin 30° = 2 \times \frac{1}{2} = 1$ so when $\theta = 30°, \sin 2\theta \neq 2\sin 2\theta$

6. $4^2 - 2^2 = 12$ (divisible by 4), $6^2 - 4^2 = 20$ (divisible by 4), $8^2 - 6^2 = 28$ (divisible by 4), so all divisible by 4

9.3 Geometric proofs

Exercise 9E

1. $\angle DFE = 180° - (90° + \frac{x°}{2}) = 90° - \frac{x°}{2}$
   $\angle DEF = 180° - x° - (90° - \frac{x°}{2}) = 90° - \frac{x°}{2}$
   $\angle DFE = \angle DEF$; hence triangle DEF is isosceles.

2. The exterior angle of a triangle is equal to the sum of the opposite two interior angles.
   $x° = \frac{x°}{2} + \frac{x°}{2}$; hence the triangle is isosceles

3. Draw radius OB. Angle OAB = angle OBA = x
   Angle OCB = angle OBC = y
   Because AOC is a straight line,
   $180° - 2x + 180° - 2y = 180°$
   $2x + 2y = 180°$
   $x + y = 90° = $ angle ABC

4. $\angle CED + \angle AEC = 180°$ (angles on a straight line)
   $\angle ABC + \angle AEC = 180°$ (cyclic quadrilateral)
   But $\angle ABC = \angle ACB$ (isosceles triangle)
   Hence $\angle ACB = \angle CED$

5. PS = QR, RS = PQ (opposite sides of a parallelogram), both triangles share side QS; hence triangles are congruent by SSS.

6. Equilateral triangles have all angles equal to 60°, which are acute not obtuse.

7. a)

   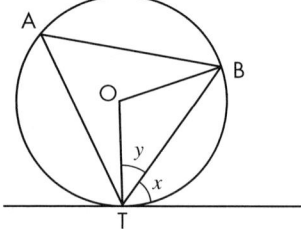

   For example, the diagram above shows a circle, centre O, with tangent PT at T.
   Angle PTO is 90° (angle between tangent and radius).
   Therefore, angle $x = 90° - y$.
   Angle TBO = angle BTO = y (isosceles triangle BOT)
   Therefore, angle BOT is $180° - 2y$ (angles in a triangle), and angle TAB = $90° - y$ (angle at circumference is half the angle at the centre subtended by arc BT).
   Hence angle PTB = angle TAB, or the angle between a tangent and a chord is equal to the angle in the alternate segment.

   b) By the alternate segment theorem $\angle TXA = \angle TYB$; hence AX is parallel to BY.

Exam-style questions

1. a) Substitute for a and b to show LHS = RHS: $(3 + 4)^2 + (3 - 4)^2 = 2(3^2 + 4^2) = 50$
   b) Expand each of the LHS and RHS to show they are algebraically the same:
   LHS $= (3 + 4)^2 + (3 - 4)^2 = 50$; RHS $= 2(3^2 + 4^2) = 50$
   c) $(a + b)^2 + (a - b)^2 = a^2 + 2ab + b^2 + a^2 - 2ab + b^2$
   $= 2a^2 + 2b^2$
   $= 2(a^2 + b^2)$

2. $(a + b)^2 - (a - b)^2 = a^2 + 2ab + b^2 - (a^2 - 2ab + b^2) = 4ab$

3. a) $7^2 - 5^2 = 24$, and $2(7 + 5) = 24$
   b) $a^2 - b^2 = (a - b)(a + b) = 2(a + b)$, because $(a - b) = 2$
   c) $a^2 - b^2 = (a - b)(a + b) = n(a + b)$, because $(a - b) = n$

4. a) $4 \times 5 \times 6 \times 7 + 1 = 841 = 29^2$
   b) Expand and simplify the LHS to the expression on the RHS:
   $(n^2 - n - 1)^2 = n^4 - n^3 - n^2 - n^3 + n^2 + n - n^2 + n + 1$
   $= n^4 - 2n^3 - n^2 + 2n + 1$
   c) $n(n - 2)(n - 1)(n + 1) + 1 = n(n^2 - 1)(n - 2) + 1$
   $= n^4 - 2n^3 - n^2 + 2n + 1 = (n^2 - n - 1)^2$ (from **b**).

5. Let the integers be n and $(n + 1)$.
   Then $n^2 + (n + 1)^2$ expands to give $n^2 + n^2 + 2n + 1$, which simplifies to $2(n^2 + n) + 1$, which is odd.

6. a) $10^2 = 100$, $1 + 8 + 27 + 64 = 100$
   b) $\left(\frac{1}{2}n(n + 1)\right)^2 = \frac{1}{4}n^2(n + 1)^2$
   c) $1 + 8 + 27 + 64 + 125 + 216 = 441$
   $\frac{1}{4} \times 36 \times 49 = 441$

7. $\angle QAT = \angle QTA$ (isosceles triangle)
   $\angle PTB = \angle QTA$ (vertically opposite angles)
   $\angle PTB = \angle PBT$ (isosceles triangle)
   Hence $\angle PBT = \angle QAT$ and PB is parallel to AQ

8. $n^3 - n = n(n^2 - 1) = n(n + 1)(n - 1)$
   $= (n - 1)(n)(n + 1)$

9.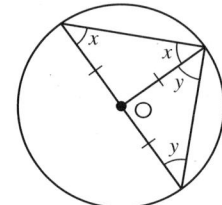

   $(180° - 2x) + (180° - 2y) = 180°$, $360° - 2x - 2y = 180°$,
   $180° = 2x + 2y$, $90° = x + y$

10. angle AOD = $2x$, angle ABD = x (angle at the circumference is half the angle at the centre). For the same reason angle ACD = x.
    Therefore, angle ABD = angle ACD.

    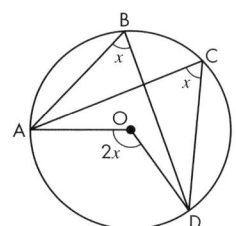

CHAPTER 10 VECTORS

10.1 Position vectors

Exercise 10A

1. a) $\begin{pmatrix} 0 \\ 2 \end{pmatrix}$ b) $\begin{pmatrix} 1 \\ 8 \end{pmatrix}$ c) $\begin{pmatrix} 6 \\ 1 \end{pmatrix}$ d) $\begin{pmatrix} 6 \\ 17 \end{pmatrix}$
   e) $\begin{pmatrix} 4 \\ -10 \end{pmatrix}$ f) $\begin{pmatrix} 18 \\ 5 \end{pmatrix}$

2. a) $-\mathbf{a}+\mathbf{b}$ b) $2\mathbf{a}+\mathbf{b}$ c) $-2\mathbf{a}-\mathbf{b}$
   d) $-\mathbf{a}-\frac{1}{2}\mathbf{b}$ e) $2\mathbf{a}-\frac{1}{2}\mathbf{b}$

3. $x = -4.5$
4. $x = -1, y = 3$
5. $\begin{pmatrix} 3 \\ 2 \end{pmatrix}$
6. $5\sqrt{2}$
7. 12.65

Exercise 10B

1. a) $\mathbf{a} = -2\begin{pmatrix} 3 \\ 1 \end{pmatrix}, \mathbf{b} = 5\begin{pmatrix} 3 \\ 1 \end{pmatrix}, \mathbf{a} = \frac{-2}{5}\mathbf{b}$, so $\mathbf{a}$ and $\mathbf{b}$ are parallel
   b) $-2 : 5$

2. $\begin{pmatrix} 1 \\ 6 \end{pmatrix}$

3. $\overrightarrow{AB} = -\overrightarrow{OA} + \overrightarrow{OB} = -\begin{pmatrix} -5 \\ 1 \end{pmatrix} + \begin{pmatrix} -3 \\ 2 \end{pmatrix} = \begin{pmatrix} 2 \\ 1 \end{pmatrix}$,
   $\overrightarrow{BC} = -\overrightarrow{OB} + \overrightarrow{OC} = -\begin{pmatrix} -3 \\ 2 \end{pmatrix} + \begin{pmatrix} 3 \\ 5 \end{pmatrix} = \begin{pmatrix} 6 \\ 3 \end{pmatrix} = 3\begin{pmatrix} 2 \\ 1 \end{pmatrix}$
   $3\overrightarrow{AB} = \overrightarrow{BC}$ and so the vectors are parallel. They also share a point, B, so they must be collinear (lie on the same straight line).

4. $\overrightarrow{DE} = -\overrightarrow{OD} + \overrightarrow{OE} = -(\mathbf{a}+\mathbf{b}) + (-2\mathbf{a}+3\mathbf{b})$
   $= (-3\mathbf{a}+2\mathbf{b})$
   $\overrightarrow{EF} = -\overrightarrow{OE} + \overrightarrow{OF} = -(-2\mathbf{a}+3\mathbf{b}) + (-11\mathbf{a}+9\mathbf{b})$
   $= (-9\mathbf{a}+6\mathbf{b}) = 3(-3\mathbf{a}+2\mathbf{b})$
   $3\overrightarrow{DE} = \overrightarrow{EF}$ and so the vectors are parallel. They also share a point, E, so they must be collinear (lie on the same straight line).

5. $\overrightarrow{BD} = \overrightarrow{BC} + \overrightarrow{CD} = \overrightarrow{BC} - \overrightarrow{AB} = 3\mathbf{a}+4\mathbf{b} - (-2\mathbf{a}+2\mathbf{b})$
   $= 5\mathbf{a}+2\mathbf{b}$
   $\overrightarrow{AE} = \overrightarrow{AD} + \overrightarrow{DE} = \overrightarrow{BC} + \overrightarrow{BD} = 3\mathbf{a}+4\mathbf{b} + 5\mathbf{a}+2\mathbf{b}$
   $= 8\mathbf{a}+6\mathbf{b}$

10.2 Solving geometric problems

Exercise 10C

1. $\overrightarrow{CE} = 3\overrightarrow{CD}$; $-9\mathbf{a}+12\mathbf{b} = 3(-3\mathbf{a}+k\mathbf{b})$, $k = 4$
2. $k = \frac{1}{5}$
3. $\overrightarrow{PS} = \frac{1}{3}(4\mathbf{a}+6\mathbf{b})$, $\overrightarrow{PT} = \frac{4}{3}(4\mathbf{a}+6\mathbf{b})$ so the vectors are parallel. They also share a point, P, and so they must be collinear.
4. 122.5°
5. 72.4°
6. $n = 4$
7. $3 : 4$
8. $\overrightarrow{WZ} = \mathbf{a}+3\mathbf{b}$ and $\overrightarrow{XY} = \mathbf{a}+3\mathbf{b}$, $\overrightarrow{YZ} = -2\mathbf{a}+2\mathbf{b}$ and $\overrightarrow{XW} = -2\mathbf{a}+2\mathbf{b}$
   WXYZ has two equal and parallel sides so is a parallelogram.

Exam-style questions

1. a) $\overrightarrow{YW} = \overrightarrow{YZ} + \overrightarrow{ZW} = 2\mathbf{a}+\mathbf{b}+\mathbf{a}+2\mathbf{b} = 3\mathbf{a}+3\mathbf{b} = 3(\mathbf{a}+\mathbf{b}) = 3\overrightarrow{XY}$
   b) $3 : 1$
   c) They lie on a straight line (they are collinear).
   d) Points are A(6, 2), B(1, 1) and C(2, −4).
   Using Pythagoras' theorem,
   $AB^2 = 26$, $BC^2 = 26$ and $AC^2 = 52$
   so $AB^2 + BC^2 = AC^2$ and hence $\angle ABC$ must be a right angle.

2. $\overrightarrow{OX} = \begin{pmatrix} -2 \\ 4 \\ 3 \end{pmatrix}$

3. $\overrightarrow{VY} = \frac{2}{3}(-\mathbf{a}+2\mathbf{b})$ and $\overrightarrow{VZ} = -\mathbf{a}+2\mathbf{b}$ $\overrightarrow{VY}$ and $\overrightarrow{VZ}$ are scalar multiples of the same vector and are therefore parallel. They share the point V so are collinear.

4. $3 : 4$

5. a) $\frac{2}{3}\mathbf{a} - \frac{1}{2}\mathbf{b}$ b) $\frac{1}{21}\mathbf{a} + \frac{2}{7}\mathbf{b}$

William Collins' dream of knowledge for all began with the publication of his first book in 1819.

A self-educated mill worker, he not only enriched millions of lives, but also founded a flourishing publishing house. Today, staying true to this spirit, Collins books are packed with inspiration, innovation and practical expertise.
They place you at the centre of a world of possibility and give you exactly what you need to explore it.

Published by Collins
An imprint of HarperCollins*Publishers*
The News Building, 1 London Bridge Street, London, SE1 9GF, UK

HarperCollins*Publishers*
Macken House, 39/40 Mayor Street Upper, Dublin 1, D01 C9W8, Ireland

Browse the complete Collins catalogue at
www.collins.co.uk

© HarperCollins*Publishers* Limited 2025

10 9 8 7 6 5 4 3 2 1

ISBN 978-0-00-873947-8

All rights reserved. No part of this publication may be reproduced, stored in a retrieval system, or transmitted in any form by any means, electronic, mechanical, photocopying, recording or otherwise, without the prior written permission of the Publisher or a licence permitting restricted copying in the United Kingdom issued by the Copyright Licensing Agency Ltd, 5th Floor, Shackleton House, 4 Battle Bridge Lane, London SE1 2HX.

British Library Cataloguing-in-Publication Data

A catalogue record for this publication is available from the British Library.

Authors: Anne Stothers, Andrew Milne
Contributing author: Trevor Senior
Publisher: Katie Sergeant
Product manager: Jennifer Hall
Editor: Julie Bond
Proofreader and answer checker: Steven Matchett
Cover designer: Amparo Barrera, Kneath Associates
Internal designer: Ken Vail Graphic Design
Illustrators: Ann Paganuzzi and Jouve India Private Limited
Typesetter: Jouve India Private Limited
Production controller: Alhady Ali
Printed in India by Multivista Global Pvt. Ltd

Acknowledgements

The publishers gratefully acknowledge the permission granted to reproduce the copyright material in this book. Every effort has been made to trace copyright holders and to obtain their permission for the use of copyright material. The publishers will gladly receive any information enabling them to rectify any error or omission at the first opportunity.

This book contains FSC™ certified paper and other controlled sources to ensure responsible forest management.

For more information visit: www.harpercollins.co.uk/green